A Study on Usage and Impact

Eddie C. Y. Kuo, Alfred Choi,
Arun Mahizhnan, Lee Wai Peng, Christina Soh

TIMES ACADEMIC PRESS

© Nanyang Technological University, Singapore, 2002

First published 2002 by
Times Academic Press
An imprint of Times Media Private Limited
(A member of the Times Publishing Group)
Times Centre, 1 New Industrial Road, Singapore 536196
for
Nanyang Technological University
31 Nanyang Link
Singapore 637718

All rights reserved. No part of this publication may be reproduced, stored in a retrieval system, or transmitted, in any form or by any means, electronic, mechanical, photocopying, recording or otherwise, without the prior permission of the publishers.

ISBN 981 210 191 8

Printed by B & Jo Enterprise Pte Ltd, Singapore.

All orders for this book should be addressed to:
                Times Media Private Limited
                Times Centre
                1 New Industrial Road
                Singapore 536196
                Fax: (65) 2889 254
                E-mail: tap@tpl.com.sg
                Online Book Store: http://www.timesone.com.sg/tap

Contents

1.	**THE SINGAPORE INTERNET PROJECT**	*1*
1.1	INTRODUCTION	*1*
1.2	THE WORLD INTERNET PROJECT (WIP)	*4*
2.	**LITERATURE REVIEW**	*6*
2.1	INTRODUCTION	*6*
2.2	USAGE AND ADOPTION	*6*
2.3	EFFECTS OF INTERNET USAGE ON TRADITIONAL MEDIA	*9*
2.4	THE INTERNET AND SOCIAL RELATIONSHIPS	*11*
2.5	DIGITAL DIVIDE	*15*
2.6	REGULATION AND POLICY	*17*
3.	**METHODOLOGY**	*19*
3.1	ADVANTAGES OF THE DESIGN	*19*
4.	**FINDINGS (ADULT SURVEY)**	*22*
4.1	SAMPLE DEMOGRAPHICS	*22*
4.1.1	POPULATION AND SAMPLING	*22*
4.1.2	DEMOGRAPHIC CHARACTERISTICS OF RESPONDENTS	*22*
4.2	USAGE PATTERNS	*24*
4.2.1	INTERNET USE	*24*

4.2.2	DEMOGRAPHICS OF USERS AND NON-USERS	25
4.2.3	INTERNET SKILLS	29
4.2.4	INTERNET ACTIVITIES	30
4.3	**E-COMMERCE**	**31**
4.3.1	DEMOGRAPHIC PROFILES	33
4.3.2	INTERNET USE AN SKILL PROFILES	34
4.3.3	GOODS AND SERVICES	35
4.3.4	CONCERNS ABOUT PURCHASING ONLINE	36
4.4	**LIFESTYLE AND WELL-BEING**	**36**
4.4.1	ACTIVITIES	37
4.4.2	SOCIAL AND FAMILY LIFE	38
4.4.3	WORK AND SCHOOL PERFORMANCE	42
4.4.4	WELL-BEING	43
4.5	**PERCEPTION OF THE INTERNET**	**46**
4.5.1	INTERNET AND POLITICS	51
4.5.2	IMPLICATIONS FOR IT PROMOTERS AND POLICY MAKERS	53
4.6	**INTERNET REGULATION**	**54**
4.6.1	CONCERNS OVER INTERNET CONTENT	54
4.6.2	WHO SHOULD BE RESPONSIBLE?	56
4.6.3	RATING SYSTEM	57
4.6.4	FILTER OPTIONS	58
4.6.5	IMPLICATIONS FOR GOVERNMENT, ISPs, CONTENT PROVIDERS AND THE PUBLIC	60
4.7	**IMPORTANCE OF INFORMATION SOURCES**	**60**
4.7.1	MASS MEDIA	60
4.7.2	INTERPERSONAL SOURCES	63
4.7.3	IMPLICATIONS FOR TRADITIONAL MASS MEDIA AND INTERPERSONAL COMMUNICATION	65

4.8	**TRUST OF INSTITUTIONS**	66
4.8.1	IMPLICATIONS FOR INSTITUTIONS	69

5. FINDINGS (STUDENT SURVEY) — 70

5.1	**SAMPLE DEMOGRAPHICS**	70
5.1.1	POPULATION AND SAMPLING	70
5.1.2	DEMOGRAPHIC CHARACTERISTICS OF RESPONDENTS	70
5.2	**INTERNET USAGE PATTERNS**	71
5.2.1	INTERNET USE	71
5.2.2	DEMOGRAPHIC OF USERS AND NON-USERS	72
5.2.3	INTERNET SKILLS	77
5.2.4	INTERNET ACTIVITIES	78
5.3	**LIFESTYLE AND WELL-BEING**	80
5.3.1	ACTIVITIES	80
5.3.2	SOCIAL AND FAMILY LIFE	81
5.3.3	SCHOOL PERFORMANCE	84
5.3.4	WELL-BEING	84
5.4	**PERCEPTIONS OF THE INTERNET**	87
5.5	**IMPORTANCE OF INFORMATION SOURCES**	92
5.5.1	MASS MEDIA	92
5.5.2	INTERPERSONAL SOURCES	95
5.6	**TRUST OF INSTITUTIONS**	96

6. SUMMARY & CONCLUSION — 100

6.1	**INTERNET ACCESS**	100
6.2	**INTERNET ACTIVITIES**	103
6.3	**ATTITUDE AND PERCEPTIONS**	104

6.4	LIFESTYLE AND WELL-BEING	105
6.5	MEDIA USE	107
6.6	INTERNET REGULATION	110
6.7	POLITICAL EMPOWERMENT	111
6.8	TRUST OF INSTITUTIONS	111
6.9	LOOKING AHEAD	112

REFERENCES *114*

ACKNOWLEDGEMENT *119*

INDEX *121*

Table of Figures

TABLE 4.1.1 :	GENDER – SAMPLE VS. POPULATION	23
TABLE 4.1.2 :	RACE – SAMPLE VS. POPULATION	23
TABLE 4.1.3 :	AGE – SAMPLE VS. POPULATION	23
TABLE 4.1.4 :	LEVEL OF EDUCATION –SAMPLE VS. POPULATION	24
TABLE 4.1.5 :	HOUSING TYPE – SAMPLE VS. POPULATION	24
TABLE 4.2.1 :	PROPORTION OF COMPUTER USERS AND NON-USERS	25
TABLE 4.2.2 :	PROPORTION OF INTERNET USERS AND NON-USERS	25
TABLE 4.2.3 :	INTERNET USERS/NON-USERS, BY GENDER	26
TABLE 4.2.4 :	INTERNET USERS/NON-USERS, BY RACE	26
TABLE 4.2.5 :	INTERNET USERS/NON-USERS, BY AGE	27
TABLE 4.2.6 :	INTERNET USERS/NON-USERS, BY MARITAL STATUS	27
TABLE 4.2.7 :	INTERNET USERS/NON-USERS, BY MONTHLY FAMILY INCOME	28
TABLE 4.2.8 :	INTERNET USERS/NON-USERS, BY HOUSING TYPE	28
TABLE 4.2.9 :	INTERNET USERS/NON-USERS, BY EDUCATION	29
TABLE 4.2.10 :	INTERNET SKILLS AMONG USERS	30
TABLE 4.2.11 :	TIME USE AT VARIOUS LOCATIONS	31
TABLE 4.2.12 :	TIME SPENT ON VARIOUS INTERNET ACTIVITIES	31
TABLE 4.3.1 :	E-COMMERCE: PURCHASERS, BROWSERS AND NON-SHOPPERS	32
TABLE 4.3.2 :	DEMOGRAPHIC PROFILES OF PURCHASERS, BROWSERS AND NON-SHOPPERS	33
TABLE 4.3.3 :	INTERNET SKILLS OF PURCHASERS, BROWSERS AND NON-SHOPPERS	34
TABLE 4.3.4 :	GOODS AND SERVICES PURCHASED AND BROWSED ON THE INTERNET	35

TABLE 4.3.5 :	CONCERNS ABOUT ONLINE PURCHASING	36
TABLE 4.4.1 :	TIME SPENT ON VARIOUS ACTIVITIES	37
TABLE 4.4.2 :	REPORTED CHANGE IN SOCIAL CONTACT WITH INTERNET USE	39
TABLE 4.4.3 :	REPORTED CHANGE IN SOCIAL INTERACTION WITH INTERNET USE	40
TABLE 4.4.4 :	FAMILY INTERACTION – "YOU AND FAMILY ARE TOO BUSY TO TALK TO EACH OTHER"	41
TABLE 4.4.5 :	FAMILY INTERACTION – "YOU AND FAMILY LISTEN TO EACH OTHER WHEN THERE IS DISAGREEMENT"	41
TABLE 4.4.6 :	FAMILY INTERACTION – "YOU AND FAMILY DO NOT REALLY UNDERSTAND WHAT EACH OTHER IS GOING THROUGH"	41
TABLE 4.4.7 :	FAMILY INTERACTION – "YOU AND FAMILY SHARE IDEAS AND OPINIONS"	42
TABLE 4.4.8 :	FAMILY INTERACTION – "YOU AND FAMILY ARE PATIENT WITH ONE ANOTHER"	42
TABLE 4.4.9 :	PRODUCTIVITY WITH INTERNET USE	43
TABLE 4.4.10 :	INTERNET USE AND OFFICE WORK AT HOME	43
TABLE 4.4.11 :	ALIENATION MEASUREMENT – "FEELING HELPLESS"	44
TABLE 4.4.12 :	ALIENATION MEASUREMENT – "AVERAGE CITIZEN CAN HAVE INFLUENCE ON GOVERNMENT"	44
TABLE 4.4.13 :	ALIENATION MEASUREMENT – "CAN CHANGE COURSE OF WORLD EVENTS"	45
TABLE 4.4.14 :	ALIENATION MEASUREMENT – "AVERAGE PEOPLE ARE GETTING WORSE"	45
TABLE 4.4.15 :	ALIENATION MEASUREMENT – "DON'T KNOW WHO TO COUNT ON"	45
TABLE 4.4.16 :	ALIENATION MEASUREMENT – "MOST PEOPLE DON'T CARE WHAT HAPPENS TO OTHERS"	46

TABLE 4.5.1	:	PERCEPTION OF INTERNET – "INTERNET IS UNIMPORTANT"	47
TABLE 4.5.2	:	PERCEPTION OF INTERNET: "INTERNET IS A USEFUL TOOL"	47
TABLE 4.5.3	:	PERCEPTION OF INTERNET – "INTERNET IS UNINTERESTING"	47
TABLE 4.5.4	:	PERCEPTION OF INTERNET – "INTERNET IS EASY TO USE"	48
TABLE 4.5.5	:	PERCEPTION OF INTERNET – "INTERNET MAKES LIFE MORE CONVENIENT"	48
TABLE 4.5.6	:	PERCEPTION OF INTERNET – "INTERNET IS A GOOD THING FOR SOCIETY"	49
TABLE 4.5.7	:	PERCEPTION OF INTERNET – "INTERNET IS USEFUL TO STAY INFORMED"	49
TABLE 4.5.8	:	PERCEPTION OF INTERNET: "INTERNET IS GREAT FOR ENTERTAINMENT PURPOSES"	50
TABLE 4.5.9	:	PERCEPTION OF INTERNET – "INTERNET MAKES BUSINESS/COMMERCIAL ACTIVITIES MORE CONVENIENT"	50
TABLE 4.5.10	:	PERCEPTION OF INTERNET – "INTERNET SKILLS ARE IMPORTANT FOR GETTING A GOOD JOB"	51
TABLE 4.5.11	:	INTERNET AND POLITICS – "PEOPLE CAN HAVE MORE POLITICAL POWER"	52
TABLE 4.5.12	:	INTERNET AND POLITICS – "PEOPLE WILL HAVE MORE SAY ABOUT WHAT THE GOVERNMENT DOES"	52
TABLE 4.5.13	:	INTERNET AND POLITICS – "PEOPLE CAN BETTER UNDERSTAND POLITICS"	53
TABLE 4.5.14	:	INTERNET AND POLITICS – "PUBLIC OFFICIALS WILL CARE MORE ABOUT WHAT PEOPLE THINK"	53
TABLE 4.6.1	:	CONCERN ABOUT UNCENSORED PORNOGRAPHIC CONTENT	55
TABLE 4.6.2	:	CONCERN ABOUT UNCENSORED POLITICAL CONTENT	55

TABLE 4.6.3	:	CONCERN ABOUT UNCENSORED RELIGIOUS CONTENT	56
TABLE 4.6.4	:	CONCERN ABOUT RACIAL CONTENT	56
TABLE 4.6.5	:	RESPONSIBILITY FOR INTERNET REGULATION	57
TABLE 4.6.6	:	NEED FOR RATING SYSTEM BY CONTENT PROVIDERS	57
TABLE 4.6.7	:	"NEED FOR RATING SYSTEM BY CONTENT PROVIDERS", BY AGE	58
TABLE 4.6.8	:	"NEED RATING SYSTEM BY CONTENT PROVIDERS", BY EDUCATION	58
TABLE 4.6.9	:	PROVISION OF FILTER OPTIONS BY ISPs	59
TABLE 4.6.10	:	"PROVISION OF FILTER OPTIONS BY ISPs", BY AGE	59
TABLE 4.6.11	:	"PROVISION OF FILTER OPTIONS BY ISPs", BY EDUCATION	59
TABLE 4.7.1	:	INTERNET AS A SOURCE OF INFORMATION	61
TABLE 4.7.2	:	TELEVISION AS A SOURCE OF INFORMATION	62
TABLE 4.7.3	:	RADIO AS A SOURCE OF INFORMATION	62
TABLE 4.7.4	:	NEWSPAPERS AS A SOURCE OF INFORMATION	62
TABLE 4.7.5	:	MAGAZINES AS A SOURCE OF INFORMATION	63
TABLE 4.7.6	:	IMPORTANCE OF VARIOUS MEDIA AS INFORMATION SOURCES	63
TABLE 4.7.7	:	COLLEAGUES AS A SOURCE OF INFORMATION	64
TABLE 4.7.8	:	FRIENDS AS A SOURCE OF INFORMATION	64
TABLE 4.7.9	:	FAMILY AND RELATIVES AS A SOURCE OF INFORMATION	64
TABLE 4.8.1	:	TRUST IN TELEVISION NEWS	66
TABLE 4.8.2	:	TRUST IN RADIO NEWS	67
TABLE 4.8.3	:	TRUST IN NEWSPAPERS' NEWS	67
TABLE 4.8.4	:	TRUST IN ONLINE NEWS	67
TABLE 4.8.5	:	TRUST IN GOVERNMENT	68
TABLE 4.8.6	:	TRUST IN BUSINESS COMPANIES/CORPORATIONS	68
TABLE 4.8.7	:	TRUST IN RELIGIOUS ORGANISATIONS	69

TABLE 5.1.1	:	GENDER DISTRIBUTION – STUDENT SAMPLE VS. POPULATION	71
TABLE 5.1.2	:	RACE – SAMPLE VS. POPULATION	71
TABLE 5.1.3	:	STREAM – SAMPLE VS. POPULATION	71
TABLE 5.2.1	:	PROPORTION OF COMPUTER USERS AND NON-USERS	72
TABLE 5.2.2	:	PROPORTION OF INTERNET USERS AND NON-USERS	72
TABLE 5.2.3	:	INTERNET USERS/NON-USERS, BY GENDER	73
TABLE 5.2.4	:	INTERNET USERS/NON-USERS, BY RACE	74
TABLE 5.2.5	:	INTERNET USERS/NON-USERS, BY RACE AT SCHOOL	74
TABLE 5.2.6	:	INTERNET USERS/NON-USERS, BY RACE AT HOME	74
TABLE 5.2.7	:	INTERNET USERS/NON-USERS, BY MONTHLY FAMILY INCOME	75
TABLE 5.2.8	:	INTERNET USERS/NON-USERS, BY HOUSING TYPE	76
TABLE 5.2.9	:	INTERNET USERS/NON-USERS, BY FATHER'S EDUCATION	76
TABLE 5.2.10	:	INTERNET USERS/NON-USERS, BY MOTHER'S EDUCATION	77
TABLE 5.2.11	:	INTERNET SKILLS AMONG USERS	78
TABLE 5.2.12	:	INTERNET ACCESS AT VARIOUS LOCATIONS	79
TABLE 5.2.13	:	TIME USE AT VARIOUS LOCATIONS	79
TABLE 5.2.14	:	TIME SPENT ON VARIOUS INTERNET ACTIVITIES	79
TABLE 5.3.1	:	TIME SPENT ON VARIOUS ACTIVITIES (HOURS PER WEEK)	80
TABLE 5.3.2	:	REPORTED CHANGE IN SOCIAL INTERACTION WITH INTERNET USE	81
TABLE 5.3.3	:	FAMILY INTERACTION – "YOU AND FAMILY ARE TOO BUSY TO TALK TO EACH OTHER"	82

TABLE 5.3.4	:	FAMILY INTERACTION – "YOU AND FAMILY LISTEN TO EACH OTHER WHEN THERE IS DISAGREEMENT"	82
TABLE 5.3.5	:	FAMILY INTERACTION – "YOU AND FAMILY DO NOT REALLY UNDERSTAND WHAT EACH OTHER IS GOING THROUGH"	83
TABLE 5.3.6	:	FAMILY INTERACTION – "YOU AND FAMILY SHARE IDEAS AND OPINIONS"	83
TABLE 5.3.7	:	FAMILY INTERACTION – "YOU AND FAMILY ARE PATIENT WITH ONE ANOTHER"	83
TABLE 5.3.8	:	REPORTED CHANGE IN PRODUCTIVITY WITH INTERNET USE	84
TABLE 5.3.9	:	INTERNET USE AND EXAMINATION RESULTS	84
TABLE 5.3.10	:	ALIENATION MEASUREMENT – "FEELING HELPLESS"	85
TABLE 5.3.11	:	ALIENATION MEASUREMENT – "AVERAGE CITIZEN CAN HAVE INFLUENCE ON GOVERNMENT"	85
TABLE 5.3.12	:	ALIENATION MEASUREMENT – "CAN CHANGE COURSE OF WORLD EVENTS"	86
TABLE 5.3.13	:	ALIENATION MEASUREMENT – "AVERAGE PEOPLE ARE GETTING WORSE"	86
TABLE 5.3.14	:	ALIENATION MEASUREMENT – "DON'T KNOW WHO TO COUNT ON"	86
TABLE 5.3.15	:	ALIENATION MEASUREMENT – "MOST PEOPLE DON'T CARE WHAT HAPPENS TO OTHERS"	87
TABLE 5.4.1	:	PERCEPTION OF INTERNET – "INTERNET IS UNIMPORTANT"	88
TABLE 5.4.2	:	PERCEPTION OF INTERNET – "INTERNET IS A USEFUL TOOL"	88
TABLE 5.4.3	:	PERCEPTION OF INTERNET – "INTERNET IS UNINTERESTING"	88
TABLE 5.4.4	:	PERCEPTION OF INTERNET – "INTERNET IS EASY TO USE"	89

TABLE 5.4.5	:	PERCEPTION OF INTERNET: "INTERNET MAKES LIFE MORE CONVENIENT"	89
TABLE 5.4.6	:	PERCEPTION OF INTERNET – "INTERNET IS A GOOD THING FOR SOCIETY"	90
TABLE 5.4.7	:	PERCEPTION OF INTERNET – "INTERNET IS USEFUL TO STAY INFORMED"	90
TABLE 5.4.8	:	THE INTERNET IS GREAT FOR ENTERTAINMENT PURPOSES	91
TABLE 5.4.9	:	PERCEPTION OF INTERNET – "INTERNET MAKES BUSINESS/COMMERCIAL ACTIVITIES MORE CONVENIENT"	91
TABLE 5.4.10	:	PERCEPTION OF INTERNET – "INTERNET SKILLS ARE IMPORTANT FOR GETTING A GOOD JOB"	91
TABLE 5.5.1	:	INTERNET AS A SOURCE OF INFORMATION	93
TABLE 5.5.2	:	TELEVISION AS A SOURCE OF INFORMATION	94
TABLE 5.5.3	:	RADIO AS A SOURCE OF INFORMATION	94
TABLE 5.5.4	:	NEWSPAPERS AS A SOURCE OF INFORMATION	94
TABLE 5.5.5	:	MAGAZINES AS A SOURCE OF INFORMATION	95
TABLE 5.5.6	:	FRIENDS AS A SOURCE OF INFORMATION	96
TABLE 5.5.7	:	FAMILY AND RELATIVES AS A SOURCE OF INFORMATION	96
TABLE 5.6.1	:	TRUST IN TELEVISION NEWS	97
TABLE 5.6.2	:	TRUST IN RADIO NEWS	97
TABLE 5.6.3	:	TRUST IN NEWSPAPERS' NEWS	97
TABLE 5.6.4	:	TRUST IN ON-LINE NEWS	98
TABLE 5.6.5	:	TRUST IN GOVERNMENT	99
TABLE 5.6.6	:	TRUST IN BUSINESS COMPANIES/CORPORATIONS	99
TABLE 5.6.7	:	TRUST IN RELIGIOUS ORGANISATIONS	99

CHAPTER 1

The Singapore Internet Project

1.1 Introduction

In a 1992 report entitled *A Vision of an Intelligent Island: IT 2000 Report*, the Singapore government set out an initial blueprint for the incorporation of information technology (IT) in Singapore. The report characterises information as "the currency of the new age" and advocates the harnessing of IT for both economic and social well-being. The government's vision for Singapore by the first decade of the new century predicts that, "Singapore, the Intelligent Island, will be among the first countries in the world with an advanced nation-wide information infrastructure. It will interconnect computers in virtually every home, office, school, and factory."

The Singapore government aims to turn the country into an information technology hub in tune with its plans to transform the society into a knowledge-based society. Thus, the government is implementing strategies to spread the use of computers in everyday life and would like all households to be linked to the Internet in the very near future.

One of the key features in the *IT 2000 Report* was the integration of information technology in education as a strategy to meet the challenges of the new millennium. In recent years, it has been evident from the various governmental efforts that Singapore has big plans in wiring up the nation and in producing a workforce that is Internet savvy. Thus, this makes schools, which are the major agents in teaching information technology (IT) skills, an important component of this overall vision. In fact, the Ministry of Education is targeting the use of IT in 30% of the school curriculum in the near future.

The pace and extent of IT growth in Singapore since the publication of the 1992 report has added further impetus to the vision of an "Intelligent Island". Yet, there is an obvious lack of knowledge of and insight into the potential impact of IT applications on business and society at large. The main reason is that the advance of IT is far outpacing the acquisition of insights into its effect. The problem, however, is not unique to Singapore but common to all countries engaged in the propagation of IT.

Scholars, futurists, policy-makers and practitioners everywhere have speculated widely on the social and economic impact of the Internet. As stated in a *Newsweek* article, "There's no turning back. Once a novelty, the Internet is now transforming how we live, think, talk and love; how we go to school, make money, see the doctor and elect our leaders." (October 11, 1999, p. 9.)

However, the evidence on the impact of the Internet so far is mainly anecdotal and speculative. This is because, firstly, the impact of a new communication technology is often incremental and cumulative. Enough individuals and organisations have to adopt the technology before it can cause a sweeping change in society. Furthermore, the development of communication technology is interrelated to other social factors. There are also critics who argue against the claim that "the Internet is causing revolutionary social changes" and suggest that it may be an overstatement.

While most agree that the Internet will bring changes to our lives, even experts do not seem to agree with one another on the directions such changes might take. Some claim that the Internet can facilitate communication and mobilisation. Optimists perceive it as a new platform for commercial activities and political disclosure as well as a tool to improve the quality of life and productivity. Some would go to the extent of saying that it is a possible cure for social inequality and discrepant development among nations.

On the other hand, the Internet has also come under fire. It has been accused of being a potential cause for information overload, physical isolation, psychological alienation, and even poor health. Some fear that it is a warm bed for lies and deceptions, a fertile ground for pornography or cybersex, a channel of neo-colonialism, addiction to the Internet and a threat to existing value systems, morality, political stability and national identity. There are also concerns over the impact of the new medium on young people as well as the possibility of them coming into contact with persons of questionable characters. Meanwhile, regulating the Internet is no easy task. It is not surprising that Singaporeans have come together to form the Parents Advisory Group for the Internet (PAGi) so as to try to educate the public about the potential pitfalls and dangers of the Internet.

The debate on the effects of the Internet is ongoing and researchers have embarked on long-term studies only recently. Large-scale studies are usually generated from Western countries such as the United States, and hardly any has been done on Asia. As Singapore enters the information age, there will be increasing penetration of the Internet at the workplace, schools and at home. It becomes imperative to study the influences of the Internet on Singaporeans as well as on the society at large.

In 1998, the UCLA's Center for Communication Policy set out to determine how interactive technology would impact upon society, especially how the Internet would change the way we live, work and play. The UCLA center was keen to collaborate with partners outside of the United States and approached and teamed up with the School of Communication Studies at Nanyang Technological University (NTU) to spearhead the Singapore arm of the project.

The Singapore Internet Project[1] (SIP) is the first nation-wide survey of Internet usage and its social impact in Singapore. Funded by the Singapore Broadcasting Authority and the Infocomm Development Authority of Singapore (IDA), its objective is to collect

---

[1] The SIP Research Team comprises the following members: Dr. Eddie C. Y. KUO, Project Director. He is the Dean & Professor, School of Communication Studies (SCS). Dr. Alfred CHOI, Principal Researcher, who is an Assoc. Prof., SCS. The Co-Researchers are (In alphabetical order): Mr. ARUN Mahizhnan, Assoc. Prof. (Adjunct), SCS, Dy Director, Institute of Policy Studies, Dr. LEE Wai Peng, Asst. Prof., SCS and Dr. Christina SOH, Assoc. Prof.; Head, SIS & Director, IMARC, Nanyang Business School. The project is funded by the Infocomm Development Authority of Singapore and the Singapore Broadcasting Authority.

empirical evidence to show the consequences, if any, of the Internet, which has been adopted rapidly by many Singaporeans (and the rest of the world) in the last few years. The research has been designed to assess the impact of the new medium on individual perceptions, attitudes and behaviours, and to identify any effects on families as well as society at large over a three-year period. It is also structured to make cross-cultural comparisons with findings from similar surveys from other countries. Such comparison will allow a better understanding of the impact of the Internet within different social, economic and political contexts.

The focus of the study includes Internet usage pattern, e-commerce activities, lifestyle and well-being, perception of the Internet, Internet regulation, importance of information sources and trust of media institutions.

Two surveys were conducted in 1999 for this study. The first survey concentrated on adults while the other survey study was administered to students. In order to allow comparisons in later years, there is much overlap in the areas of study between the student survey and adult survey. The SIP research team has identified several areas of study and collected the following baseline data for the year 1999 for both the student and the adult surveys. More details about the survey will be provided in the later sections.

1.2 The World Internet Project (WIP)

As the UCLA – NTU research collaboration gains momentum, the two teams also planned and embarked on an ambitious project that would excite and invite research teams from various countries to join the World Internet Project (WIP).

The two teams began recruiting teams from other countries in early 1999. The UCLA Center for Communication Policy served as the overall coordinator of the international projects, but with special emphasis on research teams in Europe and North America. The NTU School of Communication Studies will coordinate with the teams from Asia. The UCLA and NTU research teams, being the founding members of the WIP, worked out a general framework of the project and a platform for comparative research on how people utilize the Internet and how the interactive technology will impact individuals and societies within as well as across nations.

The WIP study will explain how the Internet is changing the world — today, tomorrow and in the future. This project is the first wide-scoped and longitudinal exploration of how life is being transformed by computers and the Internet, with year-to-year comparisons of the social and cultural changes produced as people use this new technology. This is the first time that an analysis of such broad questions about the Internet has been undertaken on a global scale.

Since the announcement in the international media of the creation of the World Internet Project in June 1999, numerous parties worldwide have expressed an interest in collaboration. In addition to the teams in Singapore and the United States, teams in Japan, China, Taiwan, Hong Kong, Italy and Sweden have joined the WIP and have collected data for their first survey. The teams met for the first time in Singapore in July 2000 during the 22nd International Association of Media and Communication Research (IAMCR) General Assembly and Scientific Conference. The World Internet Project Conference was convened the following year in Sweden in August 2001 in which new teams from India, Korea, Hungary and France also participated. More countries are expected to join the WIP in the near future. There are plans for the teams in the WIP to meet again in 2002 during the 52nd International Association of Communication (ICA) Conference in Seoul, Korea, where workshops will be organised. In 2003, Japan will play host to the WIP conference.

CHAPTER 2

Literature Review

2.1 Introduction

As research on the Internet is still a relatively new field of study, the dearth of literature is understandable. Studies that have been published have concentrated on several areas. One area of research focuses on the collation of data on the Internet user and Internet usage and considerable interest has been shown with regard to the digital divide — the factors that prevent certain people from using the Internet and the consequences of not being online. Another area involves how the Internet affects traditional media. Studies have been conducted to find out whether traditional media has been or is being displaced by the Internet. Closely associated are studies that attempt to measure the perceived credibility of this new medium. Also of interest is the social impact of the Internet, in particular on computer-mediated communication. Scholars in this area look at the possibilities of developing online communities and online friendships and the impact that this would have on offline ones. Research has also been conducted on the merging of virtuality and reality as well as the problem of identity and deception online, and its consequences on identity offline.

2.2 Usage and Adoption

Internet and User Characteristics

The popularity of the Internet has increased rapidly as the number of people using the Internet has increased exponentially with more people expected to join this phenomenon daily. According to a study conducted by NUA Ltd (2001), there were approximately 407.1 million Internet users worldwide in 2000. This represented about 6.71% of the total world population and this figure is expected to increase exponentially as countries become more developed and people more affluent.

Highly developed nations had a higher percentage of Internet users among their population than the less developed ones. For example, in Sweden, the United States and Singapore, more than 56%, 59% and 44% of the population, respectively, were Internet users. In contrast, less than 7% of the population in less developed countries such as Russia, Indonesia, India and Thailand were Internet users (Nielson Net Ratings, 2001; IMRBINT, 2001; ITU, 2001; Newsbytes Asia, 2001).

Internet users spent a significant amount of time online. In Asia, the average user spent about 12 to 14.5 hours per month online (Nielsen NetRatings, 2001). Similarly, the average Internet user in Australia spent more than six hours online per month (Nielsen NetRatings, 2001). In the United States, the average user spent more than 20 hours per month looking at Internet sites (Jupiter Media Metrix, 2001).

From the UCLA Internet Report (2000) it was found that, in general, the average Internet user was a young, white male, who lived in the urban area, with an income and education level higher than the national average. The report also indicated that a significant proportion of Americans (more than two-thirds) had some form of access to the Internet.

Lenhart's (2000) survey, which was part of the Pew Internet and American Life Project, found that more than half the adults in America did not have access to the Internet and had no plans to do so, with the elderly being more resistant and less eager than the young ones to do so. This is similar to the findings of a Harvard Kennedy School of Government study conducted in 2000, which indicated that Americans over 60 years of age were only half as likely to have ever used a computer or accessed the Internet as younger people. Lenhart (2000) also indicated that there were more online whites than blacks or Hispanics, as well as more from households with greater income. Internet penetration in rural areas was generally low due to a low level of computer usage. There was also a gender difference with women lagging behind the men.

Nevertheless, the demographics of Internet users were likely to achieve greater parity in terms of income and age, as revealed in a survey by Rainie & Packel (2001). The researchers found that the online population was becoming more like the general population in terms of demographics. This was evident not only in the United States but in other countries as well. After the initial years of development, subsequent growth in the Internet population was contributed mainly by an increase in women and minorities. For instance, recent surveys have shown that a greater proportion of women in several countries used the Internet. More than 40% of Internet users in Spain, Australia, New Zealand, Hong Kong, Singapore, Taiwan, Korea and the United States were females (Asociacion para la Investigacion de los Medios de Comuncacion, 2001; Nielsen NetRatings 2001, Korean Ministry of Information and Communication, 2001). Several Internet studies have suggested that although Internet usage is becoming more demographically mainstream (Hoffman, Kalsbeek & Lovak, 1998), it is still dominated by white males of high socio-economic status (Stempel & Hargrove, 1996; Birdsell, Muzzio, Taylor & Krane, 1996).

Motivational Factors of Internet Usage

There were several motivational factors that influenced the levels of individual Internet usage. Iivari and Igbaria (1997) in their study of workers in 81 companies in Finland found that the amount of computer experience to be the most dominant factor in determining individual usage of the Internet for participants. More specifically, favourable facilitating conditions as well as positive social factors were other important factors affecting Internet and World Wide Web usage (Chueng, Chang, & Lai, 2000).

Purposes of Internet Usage

The Internet is used for two main purposes – as a source of information and as a tool for communication. James, Wotring and Forrest's (1995) study on the social impacts of electronic bulletin boards found that about 38% of the respondents used the board for the transmission of information and education. The Internet also provided a rich source for information-gathering purposes. This finding was also supported by Maignan and Lukas' (1997) study that found that the Internet was used as a source of information. The UCLA Study (2000) showed that the Internet was used mainly for surfing, e-mail, information search for hobbies and entertainment purposes.

A survey found that 90% of Internet users in Spain used the Internet to search for information (AMIC, 2001). Information gathering in the form of reading news on the Internet was also a popular activity (UCLA Centre for Communication Policy, 2000; Athaus & Tewksbury, 2000; Rainie & Packel, 2001). Other forms of information-seeking practices included the utilisation of search engines, finding hobby information, seeking health information, work-related research, tracking of financial data and seeking religious information (Spink, Bateman & Jansen, 1999; Rainie & Packel, 2001).

The Internet was used for socialising and communication purposes as well. In their study, Maignan and Lucas (1997) found that the Internet served as a tool for communication or interpersonal exchanges. The Australian Bureau of Statistics (2001) indicated that 68% of Internet users performed emailing and chatting activities online. More specifically, the use of emailing tools was a catalyst for the continued usage of the Internet. A study by Kraut, Mukhopadhyay, Szczyapula, Kiesler and Scherlis (1998) conducted to investigate usage patterns of the Internet by various households suggested that interpersonal communication would continue to be a dominant force in residential use of the Internet. A subsequent study conducted by the same researchers had similar findings (Kraut et al., 1999). They found that email drove peoples' use of the Internet. Participants not only used email to greater effect and consistency as compared to usage for the World Wide Web, they also showed greater tendencies to continue using the Internet over a period of time (Kraut et al., 1999).

2.3 Effects of Internet Usage on Traditional Media

Some research supported the findings that increased usage of the Internet has resulted in a decrease in the time spent on other traditional media. James et al. (1995) found that Internet bulletin board users spent less time on television consumption and book reading activities. Nie

& Erbing (2000) surveyed households recruited from a random telephone sample of the U.S. population. They claim that the revolution in IT affects the social aspects of everyday life, as well as the role and use of traditional media. The study found that an increase in amount of time spent on the Internet has a corresponding decrease in the amount of time spent on traditional media.

However, other researchers suggested that although there has been a decline in the use of conventional media over the years, this could not be attributable to the emergence of the Internet. Stemp III, Hargrove & Bernt (2000) conducted a national survey to investigate whether an increase in Internet usage was correlated to the decline in the use of other conventional media. They found that the Internet did not cause the decline in the use of other media, and neither was it a strong competitor to other media. In fact, Internet users are more likely to consume other media concurrently as compared to non-Internet users. This sentiment was echoed by Schweitzer (1991) who asserted that there was not a great change in the use of the traditional news media as a result of the use of the personal computer. This shows that, while users did not abandon traditional media, they did adopt new media as news sources.

A report by the Pew Research Centre (1998) found that Americans were reading, watching and listening to the news just as often as they did two years ago. However, the type of news and the way they follow it are being fundamentally reshaped. The percentage of Americans getting news from the Internet rose from 11 million to 36 million news users, with science, health, finance, and technology news being the most popular. In addition, going online for news did not affect news consumption patterns of traditional news sources.

Researchers also suggest that Internet users tend to be heavy media users in general, although the Internet has not replaced the traditional sources (Bromley and Bowles, 1995; Kaye, B. 1998).

Credibility of Internet News Sources

Few studies have focused upon the perceived credibility of this new medium as most of the existing work has been conducted to assess the credibility of the traditional media. Several demographic characteristics associated with high use of the Internet are also related to negative perceptions of credibility. Past studies on traditional media suggest that young adults are the most likely to judge the media as credible (Westley & Severin, 1964; Carter & Greenberg, 1965) and they are also the most likely to use the Internet. Westley and Sevein (1964) compared television and newspaper credibility and discovered that the credibility that an individual accords a medium is strongly related to how often the individual uses it (Westley & Severin, 1964). Similarly, people judge their preferred medium as the most credible (Rimmer & Weaver, 1987).

In a survey conducted by NATIONAL Public Relations, a leading Canadian public relations agency, it was found that the Internet credibility gap widens with age. About 26% of Canadians over 55 years of age find Internet news coverage to be credible. The survey found that younger people and frequent Internet users are more likely to trust Internet information. However, they still found traditional news coverage to be more credible than Internet Web sites. It was found that Canadians of all ages continue to put more trust into traditional sources of information such as news coverage in newspapers, radio and television and corporate brochures than in Internet websites or chatrooms.

2.4 The Internet and Social Relationships

Technological developments have resulted in increased bandwidth, wireless portability, globalised connectivity, personalisation of websites, and collaborative filtering of information. This new technology has allowed us great conveniences. Via the Internet, we can now manage our day-to-day activities such as banking, and shopping without meeting another human being. This has led many people to become increasingly dependent on the Internet, not only as a tool to make life easier but also as a vehicle for socialisation. These developments have resulted in an increase of computer-mediated communication (CMC) significantly and decreased the proportion of communication that would be spent face-to-face. As a result, many people are spending an increasing amount of time online. For them, the Internet, in particular, CMC is not merely just another form of technology or another tool (Markham, 1998). These characteristics of CMC have also shaped the nature of social relations and networks among individuals in society (Wellman, 2000a; 2000b).

Critics of computer-mediated communication such as Kiesler et al. (1984) and Beninger (1987) believe that computer-mediated communities are unable to allow genuine and substantive relationships to flourish and that they are more likely to produce social isolation than connectivity. Calhoun (1991) feels that online interaction would lack the level of intimacy and self-disclosure that normally accompanies offline communication, hence he doubts their ability to produce meaningful social bonding.

Kraut and Lundmark (1998) attempted to measure the social involvement and psychological well-being of the participants before and after they started using the Internet. They monitored 169 people in 73 households and found that greater use of the Internet was associated with corresponding declines in family communication, local social networks and an increase in depression. Reasons suggested include the possibility that the Internet was displacing the time used for other social activities or that it was displacing the strong social ties offline. (Kraut & Lundmark, 1998).

A study by Romm and Pliskin (1999) on the social impact of emails supported the view that emails facilitated petty tyranny and the abuse of power. Using a textual analysis of email messages and in-depth studies of organisational behaviour patterns, the findings implicated emails having strong political potency, with deliberate widespread negative use of emails and for planned exploitation (Romm & Pliskin, 1999). A similar study by Garton and Wellman (1995) found that emails filtered non-verbal and context cues, causing the lack of group identity and social roles, thus encouraged uninhibited non-conforming behaviour, increased verbal insults and disagreement in communication, which could cause group polarisation and problems in reaching group consensus (Garton & Wellman, 1995).

Research consisting of laboratory experiments comparing face-to-face communication and CMC emphasised the negative aspects of CMC such as a greater amount of verbal aggression and non-conforming behaviour online (Dubrovsky, Kiesler & Sethna, 1991; Siegal et al., 1986). Such findings posited speculations that the comparative anonymity afforded by CMC made the overt expressions of hostility more acceptable in online settings (Lea et al., 1992; Spears & Lea, 1994; Zimbardo, 1969, as cited in Parks & Floyd, 1996).

However, there have also been a host of studies reporting on the positive impact that CMC has on social relationships. These studies have found that CMC serves as positive reinforcement of offline relationships. Hamman's (1998) study on American Online (AOL) users found that the use of the Internet for communication has the potential to help cement pre-existing offline bonds rather than harming users' offline social network ties. Wellman and Hampton (1999) found that relations online were meshed with that of offline relationships and that the Internet strengthened local links with neighbourhoods and households. Emails were often used to set up face-to-face meetings, thus serving as an additional communication medium for individuals (Wellman & Hampton, 1999). In addition, even though CMC has been found generally to reinforce offline social relationships, these relationships have rarely been maintained through CMC alone (Wellman, 1996). The Internet supports a variety of social ties though relationships are rarely maintained through computer mediated communication alone (Wellman & Hampton, 1999).

Venkatesh (1996) found that with the increasing usage of online technology, the potential for family activity has increased, along with the creation of new household activities, such as setting up family archives and medical histories.

Park and Floyd (1996) in their review of the literature on CMC states that claims about the hostility and impersonality of CMC have been challenged repeatedly. Research on the use of email in the workplace has shown the interpersonal side of CMC. Email has been credited as having been used for the maintenance of relationships, for socialising purposes as well as receiving emotional support via email (Feldman, 1987; Finholt & Sproull, 1990; Haythornthwaite, Wellman, & Mantei, 1984; McCormick & McCormick; 1992; Rice & Love, 1987, as cited in Parks & Floyd, 1996).

In their study of 24 online newsgroups, Parks and Floyd (1996) have discovered that many of their respondents have gone on to form personal relationships with people that they have met online. These relationships occur over time and were maintained through the use of other communication channels (e.g. postal service and telephone) and often resulted in face-to-face encounters. Thus the length of time and the degree of participation were key factors in the development of these relationships. Reid (1995) also arrived at a similar conclusion, saying that online relationship development followed a different trajectory than that of offline relationships.

Parks and Floyd (1996) believe that existing theories on the development of personal relationships have by and large ignored the settings that do not involve face-to-face interaction. They have commented that the fact that relationships that begin online rarely stay there raises questions as to our understanding of cyberspace. For many participants, cyberspace is just another meeting place. Cyberspace has increasingly been seen as a medium for social interaction. Katz and Aspden (1997) suggested that the Internet is indeed a place where friendships can develop, which often results in face-to-face meetings.

Likewise, Katz and Aspden (1997) found that the success of friendship creation appears more related to Internet experience and skills than social and personality characteristics. The Internet as a medium has the effect of de-emphasizing the importance of sociability and personality differences. They come to the conclusion that skill at using the Internet is the most important determinant of online friendship formation. This may be because those who are more skilled at using the Internet would also be more skilled at transmitting the vital social cues necessary to establish and maintain a friendship, while "newbies" would find the Internet cumbersome and difficult to use (Kraut et al., 1996).

According to Chesebro and Bonsall (1989), friendships formed in cyberspace are of a unique nature. They are not based on appearance, age, gender or race but, rather, it seems that they are based on a meeting of minds or a connection of sorts. One can chat online with a friend for years without even knowing the other person's real name. However, this does not make the friendship any less meaningful or real than those made offline.

Some scholars such as Sproull and Kielser (1991) put forth the view that CMC is a tool that can both enable and constrain communication practices and processes in the workplace. Haythomthwaite and Wellman (1998) studied the social networks of a university research group and found that the frequency of communication was associated with the closeness of work and friendship ties. They found that email was used in similar ways to face-to-face communication and that the more frequent the contact among the respondents, the more "multiplex" the tie, that is, the use of a larger number of media (including CMC) to exchange a greater variety of information.

Koku, Nazer and Wellman (2000) also found that physical proximity still mattered despite technology and that email was used as a complement to fact-to-face communication rather than as an alternative. Also the stronger the ties, the more communication occurred, both face-to-face and online.

CMC has also made scholars rethink traditional notions of community, which were often defined geographically. Online communities such as those found in chatrooms, bulletin boards and newsgroups defy geographical limits but still possess many of those characteristics of the traditional community, such as group identity and a sense of belonging. Oldenberg (1989) and Rheingold (1993) suggest that the Internet provides a place where people can experience the conviviality and sense of community that has disappeared in today's hectic pace of life. Internet communities can be seen as the epitome of Benedict Anderson's (1983) "imagined community".

2.5 Digital Divide

A general framework for the discussion of the digital divide was given by the National Telecommunications and Information Administration (1999) or NTIA, which defined the digital divide as the division between those households that have access to telephones, computers and the Internet and those that do not. The digital divide has also been defined as the technology disparity gap or inequalities in the adoption and use of information technologies (Douglas, Blank, & Hindman, 2000; Wyden, 2000; Goslee & Conte, 1998). More specifically, researchers have analysed the issue at the levels of Internet access and basic computer resources (Wylan, 2000; Novak & Hoffman, 1998).

The NTIA has found that more households are connected across all demographic groups and geographic locations, but for many of these groups, the digital divide has widened, with those in the higher income and higher education bracket more likely to have access to the Internet. The digital divide is generally structured along socio-economic divisions such as income, education, age, gender and race.

Digital Divide and Income Levels

Studies have found that a structural barrier of income existed, inhibiting the use of information technologies such as computer networks and the Internet (Douglas, Bank, & Hindman, 2000). Research has found that the higher the level of income for an individual, the more likely he or she has access to information technology or computer networks (Goslee & Conte, 1998; Hoffman & Novak, 1999; NTIA, 1999). Those with higher income levels were also more likely to own personal computers, thus having greater accessibility to means of CMC (Novak & Hoffman, 1998).

Digital Divide and Education Levels

Another structural barrier to the access of digital communication technologies was the level of education of the individual (Douglas, Bank, & Hindman, 2000). Individuals who have attained a higher level of education were more likely to have access to the Internet and other new communication technologies (NTIA report, 1999; Gosless & Conte, 1998; Novak and Hoffman, 1998).

Digital Divide and Racial Groups

As most of the existing research has been carried out in the United States, findings have centered on the disparity between the white Americans and Afro-Americans. Most studies indicate that whites, the majority, were more likely than the blacks, the minorities, to own a personal computer at home and have access to a computer and the Internet (NTIA, 1999; Hoffman & Novak, 1999; Goslee & Conte, 1998). Although Novak and Hoffman (1998) indicated that the digital gap between the whites and the blacks has widened considerably over the years, a study by the same authors a year later suggested the opposite — a decreasing gap between races — might in fact be true (Hoffman & Novak, 1999).

Digital Divide and Age

Age was another factor that might have accounted for the digital divide in society. Douglas, Blank & Hindman (2000) suggested, that age together with income and education inhibited the use of information technologies. A study by Wyden (2000) sought to raise awareness about the technology disparity gap that was affecting elderly Americans. In his findings, he highlighted the lack of basic information technologies readily available to the elderly in USA, who had to rely on senior centres for such facilities. Even so, more than half of these centres did not have computer access. The lack of available resources for the aged might have thus accounted for the fact that only 8 in 100 elderly Americans had access to information technologies (Wyden, 2000). Another interesting finding by Hoffman and Novak (1999) indicated that households with children at home were likely to have greater web usage as compared to those without children.

Digital Divide and Gender

To a lesser extent, studies have also indicated the existence of a digital divide between different genders. Hoffman and Novak (1999) found that overall levels of web access and usage were lower for women than for men. Another study by Goslee and Conte (1998) also indicated that females had lesser opportunities of accessing new communication technologies.

A recent report by the Pew Internet Research has found a surge in the number of women going online. This suggests a movement towards gender parity, although there was a gender gap in some activities. However, there was no gender difference in the use of the Internet for chatting, surfing, doing school or job related research, entertainment and e-commerce (Pew Internet Research, 2000).

2.6 Regulation and Policy

Technology is a neutral medium and can thus be used for the benefit or detriment of society. Hence there is a need for regulations and policies. One aim of research is to look into and better understand the concerns associated with the Internet so as to enable the formulation of effective regulation policies. The Berteismann Foundation Germany & Australian Broadcasting Authority (1999) have identified the listed concerns about data protection and privacy, pornography, dangers of chat rooms and racist messages as the main concerns of the people in USA, Germany, and Australia (BFG & ABA, 1999).

The study showed that people regard the control of Internet content to be an important task. However, the majority had doubts that the misuse of the Internet could be effectively monitored by police measures. Self-regulating mechanisms, such as classification of content and self-control mechanisms to select or reject content, were deemed to be more effective in countering the risks of the new media than police measures, such as monitoring of illegal contents and prosecution of those involved (BFG & ABA, 1999).

In Singapore, state authorities have yet to take a harsh stance regarding the regulation of the Internet despite active discussions about the issue in the public domain. At the moment, Internet policy and regulatory framework are just part of the overall development of cultural policy. Regulation is achieved through a combination of auto-regulated censorship and control mechanisms (Lee & Birch, 2000).

CHAPTER 3

Methodology

Designed as a longitudinal study, the SIP aims to monitor the use and potential impact of the Internet over at least three years. The SIP utilises a panel design for both the adult and the student surveys.

For the adult surveys, a representative sample of the adult population was surveyed in 1999. In the year 2000, another representative sample was drawn and the survey administered to them as well. The 2000 survey questionnaire is a revised version of the 1999 questionnaire. Respondents in the 2000 survey constitute the panel that the project will monitor for the next two years.

The students who participated in the survey will also be tracked over time. In 1999, a representative sample of Secondary One students were surveyed. The same group of students will again be surveyed in the years 2000 and 2001, when they are in Secondary Two and Three, respectively. In addition, a sample of students who were in Secondary One in the year 2000 was also surveyed.

3.1 Advantages of the Design

The design of this study has several features. First, it allows the results to be projected to the rest of the Singapore population. A random sample of 1,000 households that matched the population characteristics was selected based on the sampling frame of the Department of Statistics. The respondent from each household was chosen using a pre-established scheme that helped to limit interviewers' selection bias and to ensure that the sample would match the population.

Second, the SIP is designed as a longitudinal study to last at least three years, probably longer. Annual surveys will allow researchers to track the social impact of the Internet over time. By returning to the same it will be possible to observe changes over time and hence analyse the factors associated with such changes. This can provide a fuller picture of the relationship between Internet use and the perception, attitude and certain aspects of the behaviour. In contrast, annual cross-sectional surveys can only describe the relationship at the time of the survey and will have difficulty establishing the causal links.

The SIP collects data using two samples — an adult sample and a student sample. As the adult and student surveys share a substantial number of research areas, this allows for a comparison between the two groups.

Another feature of the study is that both Internet users and non-users were surveyed. Many Internet studies, in particular those conducted by market research firms, are interested in users only. A comparison between these two groups in terms of their demographic characteristics, beliefs and behaviour will help in our understanding of the adoption and use of the Internet and its implications. Moreover, since this is a panel study in which the same sample is surveyed over time, it will also provide us with an opportunity to observe what happens when a non-user starts to adopt the new medium and vice versa.

Finally, SIP is designed with cross-cultural comparisons in mind. It includes a set of core questions that all WIP research teams will be asking of the respondents in their respective countries.

The major findings from both the adult and the student surveys are reported in the following chapters and will be focused on the following areas:

Internet Usage Patterns — The research team is interested in finding out the amount of Internet use, where people use the Internet, what the Internet is being used for (e.g. entertainment, work, etc.) and what kind of Internet skills people have (e.g. sending e-mail, setting up a web page, etc.).

Lifestyle — The research team is tracking whether the Internet will affect people's lifestyle, including whether the new medium will change people's activities (e.g. reading, sports), interaction with family and friends, and productivity. The issue of alienation is also being addressed.

Perception of the Internet — People's perception of the Internet can be affected by their use of the Internet and their usage will in turn influence their perception of the new medium. Therefore, perception is an important area of study. Perception questions of the Internet can be divided into three categories: individual utility (e.g. ease of use, convenience, etc.), societal utility (e.g. to get a job) and the influence of Internet on politics.

Importance of Information Sources and Trust of Institutions — The last area of investigation focuses on the Internet's potential to replace traditional information sources such as television and newspapers, as well as understanding how the Internet may change the level of trust people have in various institutions.

The only difference between the student and the adult survey lie in the following two areas that were covered in the adult survey only.

E-Commerce — In addition to the general usage patterns, the popularity of e-commerce is also examined. Respondents are asked to indicate the reasons for using the Internet, to search or to purchase goods/services, and to report the types of goods/services.

Attitude on Policy and Regulation — Regulating the Internet is a daunting task. One of the challenges faced by countries around the world lies in promoting the Internet for greater economic success without sacrificing public interest. The research team attempts to find out what people think about selected undesirable content on the Internet and who should be responsible for regulating such content.

The data was collected by a professional research firm. Interviewers were sent to conduct door-to-door interviews in which one person from each selected households was interviewed. Each interview lasted approximately 20 – 25 minutes. Of those contacted, about 70% accepted and completed the interviews. To date, the research team has analysed the data collected from the 1999 survey. The rest of this book will describe the methodology and findings pertaining to this set of data.

CHAPTER 4

Findings (Adult Survey)

4.1 Sample Demographics

4.1.1 *Population and Sampling*

Only Singaporeans and permanent residents aged 18 and above were selected as respondents for the survey. A stratified random sample of 1,000 households, matching the characteristics of the Singapore population, was obtained from the Department of Statistics. Interviewers were sent to these households. Upon encountering a household which was either uncooperative or when there was no one at home at that point in time, the interviewers immediately replaced them with either their right or left neighbour. The choice of left or right was done in an alternate manner. The next stage involved choosing an adult from that particular household. Since many households consisted of more than one adult, the selection was decided upon by using a pre-determined scheme that helped minimise interviewer bias while retaining randomness. There was no refusal at the individual level. A total of 949 successful cases constituted the final sample in the present analysis.

4.1.2 *Demographic Characteristics of Respondents*

Table 4.1.1 through Table 4.1.5 compares the sample with the population in terms of gender, race, age, level of education and housing type. It was found that the sample matched the population closely in terms of gender and race. On the other two variables, the respondents in the sample were generally more educated and living in better housing categories, as compared with the population. This is not surprising because people of high socio-economic status (SES) tend to be more receptive to request for interviews.

Table 4.1.1:   Gender – Sample vs. Population

Gender	Sample (%)	Population (%)
Male	54.4	50.2
Female	45.6	49.8

Population source: Singapore Department of Statistics Website, 1998 statistics.

Table 4.1.2:   Race – Sample vs. Population

Race	Sample (%)	Population (%)
Chinese	80.9	77.0
Malay	11.0	14.0
Indian	7.3	7.6
Others	0.8	1.4

Population source: Singapore Department of Statistics Website, 1998 statistics.

Table 4.1.3:   Age – Sample vs. Population

Age	Sample (%)	Population (%)
18-24	25.4	17.4
25-34	22.0	23.1
35-44	22.3	24.7
45-54	17.5	16.4
55-64	7.7	9.2
65-74	4.3	5.7
75 & above	0.7	3.5

Population source: Singapore Department of Statistics Website, 1998 statistics.

Table 4.1.4: Level of Education –Sample vs. Population

Education Level	Sample (%)	Population (%)
Primary or below	23.3	48.3
O level /equivalent	35.7	25.5
A level	13.5	8.2
Diploma	15.3	8.0
Degree and above	12.2	10.0

Population source: Singapore Department of Statistics Website, 1998 statistics

Table 4.1.5: Housing Type – Sample vs. Population

Housing Type	Sample (%)	Population (%)
HDB 1-2 room flat	3.5	7.8
HDB 3 room flat	41.8	31.4
HDB 4 room flat	34.7	35.4
HDB 5 room flat/ HDB executive/ HUDC	20.0	25.3

Population Source: Housing and Development Board Website, 1997/98 statistics.

4.2 Usage Patterns

4.2.1 Internet Use

The study found that more than half (57.2%) of the respondents were computer users. However, only 45.7% of the respondents were Internet users (see Tables 4.2.1 and 4.2.2). "Users" in this report, referred to respondents who, at the time of the survey, said that they accessed the Internet. Former users were considered "non-users" in this context. Since the sample was representative of the population, we estimated that Internet penetration rate among adult Singaporeans, aged 18 and above, to be at about 46%.

Table 4.2.1: Proportion of Computer Users and Non-Users

	Percent (n=949)
Users	57.2
Non-Users	42.8
Total	100.0

Table 4.2.2: Proportion of Internet Users and Non-Users

	Percent (n=949)
Users	45.7
Non-Users	54.3
Total	100.0

Since the Internet is a relatively new medium, the users had only an average of 2.7 years of experience with it. When respondents were asked why they did not use or had stopped using the Internet, the top three reasons given were no time, did not know how to and no interest.

4.2.2 *Demographics of Users and Non-Users*

This study found significant differences between users and non-users in terms of their demographic characteristics. Overall, Internet users were more likely to be male, younger, single and had better socio-economic status in terms of income, housing and education.

Gender. The study found significant gender difference among the respondents. Among male respondents, there were more users (56%) than non-users (44%). However, the reverse was true for female respondents — 33.5% were users, while 66.5% were non-users (see Table 4.2.3)

Table 4.2.3: Internet Users/Non-Users, by Gender

	Users Percent	Non-Users Percent	Total (n=949) Freq	Percent
Male	56.0	44.0	516	100.0
Female	33.5	66.5	433	100.0

(Chi-square = 48.1, d.f. = 1, p = 0.0.)

Race. As Table 4.2.4 indicates, the study found no racial differences among the respondents. The proportions of users and non-users for Chinese, Malays and Indians were comparable.

Table 4.2.4: Internet Users/Non-Users, by Race

	Users Percent	Non-Users Percent	Total (n=949) Freq	Percent
Chinese	45.4	54.6	768	100.0
Malay	43.3	56.7	104	100.0
Indian	49.3	50.7	69	100.0
Others	75.0	25.0	8	100.0

(Chi-square test indicated no significant difference.)

Age. Internet users were found to be younger. Table 4.2.5 shows that as age increased the percentage of users decreased, while the reverse was true among non-users. We found a significantly higher number of users among young people.

Table 4.2.5: Internet Users/Non-Users, by Age

	Users Percent	Non-Users Percent	Total (n=949) Freq	Percent
18-24	77.6	22.4	241	100.0
25-34	50.7	49.3	209	100.0
35-44	43.4	56.6	212	100.0
45-54	21.1	78.9	166	100.0
55-64	16.4	83.6	73	100.0
65-74	4.9	95.1	41	100.0
75 and up	0	100.0	7	100.0

(Chi-square = 200.5, d.f. = 6, p = 0.0.)

Marital status. As shown in Table 4.2.6, the percentage of users (72.1%) was higher than non-users (27.9%) among single respondents. However, the trend was reversed among married people — we found more non-users (71.4%) than users (28.6%).

Table 4.2.6: Internet Users/Non-Users, by Marital Status

	Users Percent	Non-Users Percent	Total (n=949) Freq	Percent
Single	72.1	27.9	376	100.0
Married	28.6	71.4	562	100.0
Others	18.2	81.8	11	100.0

(Chi-square = 174.6, d.f. = 2, p = 0.0.)

Income. The findings clearly indicates that users were more affluent than non-users. Table 4.2.7 shows the distribution of respondents' family income[2]. In the lowest three income groups, namely $2,000-or-less, $2,001-$3,000 and $3,001-$4,000, there were more non-users than

---

[2] Respondents who indicated that they did not know the amount and those who refused to answer were omitted from the analysis.

users. However, we found the opposite among the higher income brackets, where there were more users than non-users.

Table 4.2.7: Internet Users/Non-Users, by Monthly Family Income

	Users Percent	Non-Users Percent	Total (n=904)	
			Freq	Percent
$2000 or less	22.3	77.7	175	100.0
$2001 - $3000	37.7	62.3	199	100.0
$3001 - $4000	47.8	52.2	228	100.0
$4001 - $5000	57.4	42.6	155	100.0
$5001 - $6000	59.7	40.3	77	100.0
$6001 - $7000	82.4	17.6	34	100.0
Above $7000	86.1	13.9	36	100.0

(Chi-square = 100.8, d.f. = 6, p = 0.0.)

Housing. The observation that Internet users were more affluent than non-users was further confirmed by housing types. Table 4.2.8 presents the distribution of housing type of the respondents, respectively. Again, the percentages of non-users as opposed to non-users were higher in the HDB 1-2 room and HDB 3-room housing types. The proportions were comparable among respondents who lived in HDB 4-room flats — there were 49% users and 51% non-users. We saw more users than non-users in HDB 5-room flats or better housing.

Table 4.2.8: Internet Users/Non-Users, by Housing Type

	Users Percent	Non-Users Percent	Total (n=949)	
			Freq	Percent
HDB 1 or 2-Room	35.5	64.5	31	100.0
HDB 3-Room	36.3	63.7	372	100.0
HDB 4-Room	48.9	51.1	309	100.0
HDB 5-Room/ HDB Executive/ HUDC	59.0	41.0	178	100.0
Private	54.2	45.8	59	100.0

(Chi-square = 30.2, d.f. = 4, p = 0.0.)

Education. Internet usage was related to education. We found more non-users than users among respondents with lower levels of education. In fact, 95.5% of those with primary or below education were non-users. The percentage of non-users was lower among respondents with O-level or equivalent education, but it was still at 61.1%. The trend reversed as education hit A-levels, which had 62.5% users and 37.5% non-users. There were also more users than non-users among respondents with Diplomas (78.6% vs. 21.4%) and University and above (84.5% vs. 15.5%) education.

Table 4.2.9: Internet Users/Non-Users, by Education

	Users Percent	Non-Users Percent	Total (n=949) Freq	Percent
Primary or below	4.5	95.5	221	100.0
'O' level or equivalent	38.9	61.1	339	100.0
A level	62.5	37.5	128	100.0
Diploma	78.6	21.4	145	100.0
Degree and above	84.5	15.5	116	100.0

(Chi-square = 305.4, d.f. = 4, p = 0.0.)

4.2.3 Internet Skills

Internet users were asked to indicate their ability to handle six kinds of Internet skills — sending e-mail, print and save Internet files, use search engines, download information (including graphics and software programmes), participate in online chat groups and set up a web page. A total of 92.2% Internet users were able to perform three or more skills. Only 1.6% of users could carry out one or none of the six listed items.

The findings showed that most Internet users knew how to send e-mail (82%), print and save files (89.2%), and were able to download information (73.6%). Also, 70.5% could participate in online discussions and 69% were able to use search engines. Compared to other Internet skills, setting up a web site was not as widely known — only 45.7% of the Internet users were able to do it (see Table 4.2.10). Overall, Internet users seemed to be quite competent at manoeuvring their way in cyberspace.

Table 4.2.10: Internet Skills among Users

	Percent (n=434)		Percent (n=434)
E-mail		Online chat	
Yes	82.0	Yes	70.5
No	18.0	No	29.5
Print/save		Download[a]	
Yes	89.2	Yes	73.6
No	10.8	No	26.4
Search engines		Set up web page	
Yes	69.0	Yes	45.7
No	31.0	No	54.3

a. Download includes retrieving information, graphics, software programmes, and other materials from the Internet.

4.2.4 Internet Activities

This section described the locations where respondents typically logged on to the Internet and how much time they spent on various Internet activities. As shown in Table 4.2.11 Internet users logged on for an average of 10.6 hours per week — of which 5.2 hours were from home and another 4.9 hours from school/work. Further analysis showed that 45.6% of users had access to the Internet both at home and at work/school. In addition, 35.9% had logged on from home, but not from work/school; and another 16.1% had chalked up some Internet time from work/school, but not from home. A minority of 2.3% users reported that they did not access the Internet from home or school/work. Overall, this confirmed the prevalence of home Internet use — about 80% of the users had home access, with an average of more than 5 hours per week.

Table 4.2.11: Time Use at Various Locations

	Average Hours Per Week
Home	5.2
School/work	4.9
Others	0.5
Total	10.6

As presented in Table 4.2.12, among the six categories of Internet activities, users spent most of their Internet time on e-mail — an average of 3.2 hours per week. They also put in about 2.3 and 1.8 hours per week searching for information related to work/school and for personal reasons, respectively. Taken together, info-seek becomes the most predominant activity, exceeding the time users spent on the Internet. Online entertainment and online chat each took up 1.7 and 1.6 hours per week.

Table 4.2.12: Time Spent on Various Internet Activities

	Average Hours Per Week
E-mail	3.2
Online discussion/chat	1.6
Entertainment	1.7
Info seek — work/school	2.3
Info seek — personal	1.8
Transaction	0.4
Total	11.0

4.3 E-Commerce

Only Internet users among the adult sample were asked the e-commerce questions. Internet users could be categorised into three groups with regards to e-commerce (see Table 4.3.1):

purchasers — those who had transacted over the Internet by buying goods or services. Only about 10.6% of Internet users fell into this category;

browsers — those who had searched for information for goods and services online, but had not made any purchases (45.4%); and

non-shoppers — those who had neither purchased nor browsed online (44%). The distinction was useful because each group had a different demographic and Internet skill profile, as well as different motivations and concerns about e-commerce.

Table 4.3.1: E-Commerce: Purchasers, Browsers and Non-Shoppers

	Percent (n=434)
Purchasers	10.6
Browsers	45.4
Non-shoppers	44.0
Total	100.0

The proportion of purchasers was low compared to the United States and Europe. While the percentage of the population that was online was about the same for the United States and Singapore (about 40%), a much higher percentage of U.S. Internet users had made online purchases. Recent studies estimated that between 28%[3] and 32%[4] of U.S. Internet users have made online purchases, as compared to about 10% in Singapore. The percentage of European households that were online was much lower (18%) than that in Singapore, but of these, about 30% had made online purchases.[5]

---

3 "Intelliquest Study Shows 83m US Internet Users and 56 Million Online Shoppers" http://www.intelliquest.com/press.
4 "One-Third of Internet Users Have Made Online Purchases" http://cyberatlas.Internet.com/big_picture/demographics/article.
5 "Forrester Research: Europeans Slow to Shop Online" http://www.forrester.com/ER/Research/Report/Excerpt/.

Table 4.3.2: Demographic Profiles of Purchasers, Browsers and Non-Shoppers

		% of Respondents in Each Category		
		Purchasers (n=46)	Browsers (n=197)	Non-Shoppers (n=191)
Gender	Male	76.1	64.0	67.0
	Female	23.9	36.0	33.0
Age	18-24	26.1	43.7	46.6
	25-34	32.6	26.9	19.9
	35-44	30.4	19.8	20.4
	45-74	10.9	9.6	13.1
Marital Status	Single	41.3	62.9	67.0
	Married	58.7	36.5	32.5
Income	Less than $2k	6.5	8.1	11.4
	$2k to $4k	37.0	35.5	54.6
	$4k to $6k	30.4	36.6	28.6
	Above $6k	26.0	19.9	5.4
Education	Below Pr 6	2.2	1.0	3.7
	O Level	30.4	25.9	35.1
	A Level	8.7	16.2	23.0
	Diploma	23.9	30.5	22.5
	Degree	34.8	26.4	15.7

4.3.1 Demographic Profiles

This set of data shows that purchasers were more likely to be male (76%), between 25 and 44 years old (63%), married (59%), had higher family income (26% had above $6,000 per month) and were more educated (35% had at least a degree). Details could be found in Table 4.3.2.

In comparison to purchasers, browsers and non-shoppers seemed to be younger (43.7% and 46.6%, respectively, were between 18- and 24-years old) and, hence, were also more likely to be single (62.9% and 67%, respectively).

There were, however, demographic differences between browsers and non-shoppers. Browsers were more likely to be affluent (19.9%

compared to 5.4% had monthly family income of above $6,000) and more highly educated (26.4% compared to 15.7% had at least a degree) than non-shoppers.

Overall, education and income were clearly correlated with consumer related e-commerce. This was true also in the United States. However, the United States, with its longer commercial Internet experience, has seen a steady growth in the number of women shoppers and browsers.

4.3.2 Internet Use and Skill Profiles

Of all three categories, purchasers spent the most time on the Internet — an average of 12.8 hours per week. Browsers were close behind, with an average of 12.0 hours per week, while the average for non-shoppers was a relatively low 8.5 hours per week.

The skill levels were closely correlated to the hours of use. Purchasers and browsers had similar skill profiles. Using a list of six common online tasks, purchasers and browsers were able to perform an average of 5.3 and 5.0 tasks, respectively. Non-shoppers were able to perform an average of 4.4 tasks.

Table 4.3.3: Internet skills of Purchasers, Browsers and Non-Shoppers *(cont'd)*

Online Tasks	% of Internet Users who were able to Perform the Tasks		
	Purchasers (n=46)	Browsers (n=197)	Non-Shoppers (n=191)
E-mail	100.0	99.5	97.4
Use Internet Search Engines	97.8	97.5	81.2
Print or Save Web Info	97.8	97.5	93.7
Download software, photos, audio, videos	93.5	87.8	71.7
Online Chat	80.4	71.1	70.2
Set up webpage	58.7	50.3	30.4

4.3.3 Goods and Services

The goods and services most commonly purchased online were computer-related products, books/CDs/records, movie/concert tickets and travel reservations (see Table 4.3.4). Recent surveys of the U.S. and European online purchases showed the same top categories of goods and services.[6] The top three reasons that purchasers had for buying online were:

(1) ease of finding specific products and services;
(2) comprehensive Information about products and services; and
(3) products and services not available elsewhere.

The proportion of Internet users who searched for information on goods and services online were much higher than the proportion that did only purchase. Interestingly, while travel services and books/CDs/records were among the most commonly browsed items, education-related products and services was the top category for Singapore browsers. This is in line with the Singaporean interest in education and suggests that a market may exist for online delivery of educational services. The top three reasons that browsers had for browsing online were:

(1) 24-hour availability;
(2) Convenient and saves time; and
(3) Ease of finding specific products or services.

Table 4.3.4: Goods and Services Purchased and Browsed on the Internet

	Purchasers (n=46) Percent	Browsers (n=197) Percent
Computer Related	39.6	11.9
Books/ CDs/ Records	20.8	14.9
Movies/ Concerts	12.5	1.5
Travel	12.5	12.4
Education	2.1	17.5

---

6 "Results of GVU's 10th World Wide Web User Survey" http://www.gvu.gatech.edu.user_surveys.

4.3.4 Concerns about Purchasing Online

Although the number of online purchasers was small, more than half of Internet users (54%) said that they had no concerns making purchases online. Among those that expressed concerns, the top three concerns were security of credit card information, privacy of personal data, and the difficulty of assessing quality of goods and services online (see Table 4.3.5).

In the UK, Internet users also cited security risks as their top barrier to online purchases[7]. However, recent studies of U.S. Internet users seemed to suggest that concerns about product pricing and service levels (in particular delivery of goods) had overtaken credit card and privacy issues as the top barriers to online purchases[8].

Table 4.3.5: Concerns about Online Purchasing

	Purchaser	Browser	Non-Shopper
Security of credit card info	4.1	4.1	3.7
Privacy of personal data	4.0	4.1	3.7
Difficult to assess quality	3.8	4.1	3.9
Ease of returning products	3.7	4.0	3.9
Shipping charges	3.8	3.5	3.8
Too long to receive product	3.5	3.6	3.2
Delivery of damaged product	3.8	3.9	3.9
Lack of face-to-face contact	3.8	3.3	3.9

(Average rating on a 5-point scale, where 1 is "not concerned at all" and 5 is "extremely concerned".)

4.4 Lifestyle and Well-Being

The issue of Internet's impact on people's lives and well-being is a common concern and has been widely discussed. However, the answer

---

7 "Durlacher Research Ltd: UK Ecommerce demands Are Not Met" http://www.durlacher.com/fr-research.htm.
8 "One-Third of Internet Users Have Made Online Purchases" http://cyberatlas.Internet.com/big_picture/demographics/article.

has been inconclusive so far. One of the objectives of this study was to see if Internet use would affect how people spent their time, their interaction with friends and family, work and school performances, and their well being.

4.4.1 Activities

Table 4.4.1 lists how much time respondents spent on 12 different activities. There were significant differences between users and non-users for all the activities, except reading newspapers and listening to radio. Compared to non-users, users reported spending a lot more time reading book and magazines, playing video/computer games, listening to CDs, MDs, tapes and records, going to the cinemas, talking on the telephone, socialising with friends and exercising. However, non-users spent more time watching television and interacting with family members.

This should not come as a surprise as Internet users were generally younger, better educated and more likely to be single. Their demographics could explain the greater engagement with books, magazines, video/computer games, cinema and friends. On the other hand, the fact that the non-users tended to be less educated, older and married could also shed some light on why they watched more television and tended to interact with family members.

Table 4.4.1: Time Spent on Various Activities

	Average Hours Spent Per Week	
	Users	Non-Users
Books	6.9	2.1
Video/PC Games	3.3	1.5
Tapes/CDs/MDs/Records	4.5	2.8
Newspapers	5.6	5.7
Magazines	3.5	2.2
Radio	8.7	9.9
Telephone	7.5	6.0
Television	13.0	16.7
Cinema	1.8	0.9
Face-to-face: Friends	9.3	6.4

Table 4.4.1: Time Spent on Various Activities (cont'd)

	Average Hours Spent Per Week	
	Users	Non-Users
Face-to-face: Family	19.1	27.6
Exercise/Sports	3.4	2.2

T-tests showed significant difference (at alpha = 0.05 level) between users and non-users in all of the activities, except time spent on newspapers and radio.

4.4.2 Social and Family Life

Internet users were asked to report how much the Internet had changed their contact with four different types of people: those who shared with them similar hobbies/activities, political interests, religions, and professions. Overall, the percentage of respondents who said the Internet had increased their contacts out-numbered those who claimed a drop (see Table 4.4.2).

About 45% of the users said that the Internet had helped in the development of their contact with people in their professions, although 47% claimed no change and another 7.8% said it had dwindled. With people whom Internet users shared hobbies/activities, 41% indicated an increase, 42.2% said no change and 16.8% stated a reduction.

Most users said that the Internet had not altered the amount of contact they had with people of similar political interests (64.7%). Furthermore, the number of people who reported an increase (18%) was about the same as those who indicated a decrease (17.3%).

Again, most users said that the Internet had not altered the degree of their interaction with people who shared their religion (61.3%). However, there were more people who said the Internet had increased (23.2%) their religion contacts than who stated otherwise (15.4%).

Table 4.4.2: Reported Change in Social Contact with Internet Use

	Percent (n=434)
Contact with people who share hobbies/activities	
Less	16.8
No change	42.2
More	41.0
Contact with people who share political interests	
Less	17.3
No change	64.7
More	18.0
Contact with people who share religion	
Less	15.4
No change	61.3
More	23.2
Contact with people in your profession	
Less	7.8
No change	47.0
More	45.1

In order to gauge social interaction better, all Internet users were asked about their interaction with friends and family (see Table 4.4.3). The results revealed a good sign for friendship, with about half (48.6%) saying they had made more friends (as compared to 13.4% users who said they had fewer). Although most users said that Internet use had not affected the time they spent with close friends, 21% still said it had caused a drop and another 19.4% reported an increase.

Unlike friendship, the Internet's impact on family life was less rosy — 24.2% users said that the new medium had reduced the time they spent with family, as opposed to about 17% of the users who said they had experienced an increase. At this time, close to 60% of users reported no change in the time spent with family.

Table 4.4.3: Reported Change in Social Interaction with Internet Use

	Percent (n=434)
Have more friends	
Less	13.4
No change	38.0
More	48.6
Time spent with close friends	
Less	21.0
No change	59.7
More	19.4
Time spent with family	
Less	24.2
No change	58.8
More	17.1

To further investigate the impact of the Internet on family life, the researchers compared the quality of family interaction between users and non-users. More than 52% of the respondents thought they had time to talk to their families. About 25% of the respondents thought otherwise (see Table 4.4.4). About half of them also felt they had time to listen to family members when there was disagreement and less than a quarter said otherwise (see Table 4.4.5). Also, 23% to 32% of the respondents believed that their families knew what they were going through. About 38% to 48% of the respondents said their families did not know of their problems (see Table 4.4.6). More than 55% of the respondents said that they shared ideas and opinions with their families. However, 14% to 19% of them said the opposite (see Table 4.4.7). Finally, more than 56% of the respondents said they and their family members were patient with each other. People who disagreed ranged from 13% to 14% (see Table 4.4.8).

In general, statistical tests showed no significant difference between Internet users and non-users. This set of questions indicated that the use of Internet made no difference to the quality of family interaction.

Table 4.4.4: Family Interaction – "You and Family are Too Busy to Talk to Each Other"

	Users (n=432) Percent	Non-Users (n=513) Percent
Disagree	60.6	52.0
Neutral	19.7	23.4
Agree	19.7	24.6
Total	100.0	100.0

(Chi-square test indicated no significant difference between users and non-users.)

Table 4.4.5: Family Interaction – "You and Family Listen to Each Other When There is Disagreement"

	Users (n=432) Percent	Non-Users (n=513) Percent
Disagree	21.1	21.6
Neutral	29.2	27.9
Agree	49.8	50.5
Total	100.0	100.0

(Chi-square test indicated no significant differences between users and non-users.)

Table 4.4.6: Family Interaction – "You and Family Do Not Really Understand What Each Other is Going Through"

	Users (n=432) Percent	Non-Users (n=513) Percent
Disagree	48.4	38.2
Neutral	28.7	30.0
Agree	22.9	31.8
Total	100.0	100.0

(Chi-square test indicated no significant differences between users and non-users.)

Table 4.4.7: Family Interaction – "You and Family Share Ideas and Opinions"

	Users (n=432) Percent	Non-Users (n=513) Percent
Disagree	14.1	18.9
Neutral	23.4	25.5
Agree	62.5	55.6
Total	100.0	100.0

(Chi-square test indicated no significant difference between users and non-users.)

Table 4.4.8: Family Interaction – "You and Family are Patient with One Another"

	Users (n=432) Percent	Non-Users (n=513) Percent
Disagree	13.4	13.6
Neutral	22.5	30.4
Agree	64.1	55.9
Total	100.0	100.0

(Chi-square test indicated no significant differences between users and non-users.)

4.4.3 Work and School Performance

As shown in Table 4.4.9, Internet users reported increased productivity with the use of the Internet. In fact, 69.6% users reported so. Only 12.2% users replied that there was a decrease in productivity with Internet use.

Table 4.4.9: Productivity with Internet Use

	Percent (n=434)
Less	12.2
No Change	18.2
More	69.6
Total	100.0

Productivity aside, researchers were also interested in knowing how the use of Internet might have affected work. This turned out to be a small percentage. Only 13.1% of the respondents said that with the use of Internet, they had been bringing work home at night. Among those who brought work home, 72% did so 3 or fewer days per week.

In addition, less than 9% said that with the use of Internet, they worked from home during normal office hours. Again, among those who did work from home, 85.7% said they did so 3 or fewer days per week (see table 4.4.10).

Table 4.4.10: Internet Use and Office Work at Home

	Bring Work Home (n=244) Percent	Work from Home (n=244) Percent
Yes	13.1	8.6
No	86.9	91.4
Total	100.0	100.0

4.4.4 Well-Being

While the political and business communities were trying to harness the power of Internet for growth and development, one area of social concern was the influence of the Internet on people's emotional well-being. Six items were used to measure the level of alienation, comprising: if people felt helpless, if they thought they could influence the government, if they believed they could change things in the world, if they thought life was getting worse, if they could count on others, and if others cared. Results are presented in Tables 4.4.11 through 4.4.16.

There was no difference between users and non-users for half of the items. However, there were significant differences for the other three items. In fact, the statistics showed there were more non-users than users who felt helpless, thought that they could not count on others, and that others did not care.

Table 4.4.11: Alienation Measurement – "Feeling Helpless"

	Users (n=434) Percent	Non-Users (n=514) Percent
Disagree	37.6	23.2
Neutral	23.7	32.1
Agree	38.7	44.7
Total	100.0	100.0

(Chi-square = 6.8, d.f. = 2, p = 0.0.)

Table 4.4.12: Alienation Measurement – "Average Citizen Can Have Influence on Government"

	Users (n=434) Percent	Non-Users (n=514) Percent
Disagree	32.9	30.4
Neutral	34.8	39.5
Agree	32.3	30.2
Total	100.0	100.0

(Chi-square test indicated no significant difference between users and non-users.)

Table 4.4.13: Alienation Measurement – "Can Change Course of World Events"

	Users (n=434) Percent	Non-Users (n=514) Percent
Disagree	27.6	28.4
Neutral	30.2	38.1
Agree	42.2	33.5
Total	100.0	100.0

(Chi-square test indicated no significant difference between users and non-users.)

Table 4.4.14: Alienation Measurement – "Average People are Getting Worse"

	Users (n=434) Percent	Non-Users (n=514) Percent
Disagree	25.6	16.5
Neutral	38.0	43.8
Agree	36.4	39.7
Total	100.0	100.0

(Chi-square test indicated no significant difference between users and non-users.)

Table 4.4.15: Alienation Measurement – "Don't Know Who to Count On"

	Users (n=434) Percent	Non-Users (n=514) Percent
Disagree	25.1	11.9
Neutral	35.5	31.9
Agree	39.4	56.2
Total	100.0	100.0

(Chi-square = 26.9, d.f. = 2, p = 0.0.)

Table 4.4.16: Alienation Measurement – "Most People Don't Care What Happens to Others"

	Users (n=434) Percent	Non-Users (n=514) Percent
Disagree	32.0	14.0
Neutral	25.4	28.8
Agree	42.6	57.2
Total	100.0	100.0

(Chi-square = 15.2, d.f. = 2, p = 0.0.)

4.5 Perception of the Internet

People's perception of the Internet is an important area of study, as it helps explain Internet usage and impact. If people view the Internet in a positive light, they may start or continue to use the new medium. If they see that the Internet has negative attributes, they may avoid it or terminate usage. In return, the use of the Internet may also affect perception. For example, when people become familiar with the Internet, they may modify their initial belief that the medium is difficult. Since perception is related to usage, and usage to impact, it is, therefore, included in this survey.

Respondents were asked what they thought about the Internet in relation to themselves. For example, respondents reported if they thought the Internet was important to them — not to others or society at large, but to themselves. Overall, Internet users' perception of the new medium was more positive than non-users'. They were more likely to say that, to them, the Internet was important (82.7%), useful (79.3%), interesting (74%), easy to use (83.6%) and made life more convenient (78.8%). In stark contrast, non-users were not so ready to agree. The proportions of non-users who agreed to these items were much lower: important (28.3%), useful (33%), interesting (26.6%), easy to use (28.5%) and made life more convenient (33%). Although a good number of non-users took the neutral stance, those who disagreed with the benefits of the Internet ranged anywhere from 27.8% (that the Internet made life more convenient) to 44.5% (that the Internet was important). Breakdowns of the responses can be found in Table 4.5.1 through Table 4.5.5.

Table 4.5.1: Perception of Internet – "Internet is Unimportant"

	Users (n=434) Percent	Non-Users (n=515) Percent
Disagree	82.7	28.3
Neutral	7.8	27.2
Agree	9.4	44.5
Total	100.0	100.0

(Chi-square = 88.2, d.f. = 2, p = 0.0.)

Table 4.5.2: Perception of Internet: "Internet is a Useful Tool"

	Users (n=434) Percent	Non-Users (n=515) Percent
Disagree	11.1	31.3
Neutral	9.7	35.7
Agree	79.3	33.0
Total	100.0	100.0

(Chi-square = 25.5, d.f. = 2, p = 0.0.)

Table 4.5.3: Perception of Internet – "Internet is Uninteresting"

	Users (n=434) Percent	Non-Users (n=515) Percent
Disagree	74.0	26.6
Neutral	15.2	31.3
Agree	10.8	42.1
Total	100.0	100.0

(Chi-square = 62.6, d.f. = 2, p = 0.0.)

Table 4.5.4: Perception of Internet – "Internet is Easy to Use"

	Users (n=434) Percent	Non-Users (n=515) Percent
Disagree	6.0	29.3
Neutral	10.4	42.1
Agree	83.6	28.5
Total	100.0	100.0

(Chi-square = 50.1, d.f. = 2, p = 0.0.)

Table 4.5.5: Perception of Internet – "Internet Makes Life More Convenient"

	Users (n=434) Percent	Non-Users (n=515) Percent
Disagree	5.5	27.8
Neutral	15.7	39.2
Agree	78.8	33.0
Total	100.0	100.0

(Chi-square = 26.5, d.f. = 2, p = 0.0.)

As expected, the perception of the Internet was different between users and non-users, with users being more positive. Further analysis showed that experience with the Internet was positively correlated with perception. People who had used the Internet longer were more positive toward the Internet ($r = 0.4$, $p = 0.0$), and individuals who spent more time per week on the Internet were also more positive ($r = 0.2$, $p = 0.0$) — although the latter relationship was weak. These findings suggest a positive and encouraging trend toward the future growth and utilisation of the Internet in Singapore.

In addition, respondents were asked if the Internet was useful to stay informed, convenient for business activities, good for society, great for entertainment and important for people to get jobs. Overall, again, users were found to be more positive than non-users. However, unlike the above segment on perceived individual utility, the differences between users and non-users in their perception of its social utility were narrower. (Details of the percentages can be found in Table 4.5.6 through Table 4.5.10.)

Close to 80% of users said that the Internet was useful to enable individuals to stay informed, while 70.3% of non-users agreed. Relatively high proportions of users (74%) and non-users (63.7%) said that the Internet was a good thing for society.

As for the other three items, the discrepancies between users and non-users were larger. Users were more ready to say that the Internet was convenient for business activities (79%), great for entertainment purposes (72.1%) and important for people to get good jobs (61.8%). However, the proportions of non-users who agreed were 48.7%, 49.9%, and 38.6%, respectively. A substantial number of non-users remained "neutral" as they could not decide if the Internet was beneficial to business activities (40.4%), entertainment (34.2%) and getting jobs (45%).

Table 4.5.6: Perception of Internet – "Internet is a Good Thing for Society"

	Users (n=434) Percent	Non-Users (n=515) Percent
Disagree	11.8	11.5
Neutral	14.3	24.9
Agree	74.0	63.7
Total	100.0	100.0

(Chi-square test indicated no significant difference between users and non-users.)

Table 4.5.7: Perception of Internet – "Internet is Useful to Stay Informed"

	Users (n=434) Percent	Non-Users (n=515) Percent
Disagree	7.4	9.3
Neutral	9.7	20.4

Table 4.5.7: Perception of Internet – "Internet is Useful to Stay Informed" (cont'd)

	Users (n=434) Percent	Non-Users (n=515) Percent
Agree	79.9	70.3
Total	100.0	100.0

(Chi-square test indicated no significant difference between users and non-users.)

Table 4.5.8: Perception of Internet: "Internet is Great for Entertainment Purposes"

	Users (n=434) Percent	Non-Users (n=515) Percent
Disagree	9.2	15.9
Neutral	18.7	34.2
Agree	72.1	49.9
Total	100.0	100.0

(Chi-square = 9.5, d.f. = 2, $p = 0.0$.)

Table 4.5.9: Perception of Internet – "Internet Makes Business/Commercial Activities More Convenient"

	Users (n=434) Percent	Non-Users (n=515) Percent
Disagree	6.2	10.9
Neutral	14.8	40.4
Agree	79.0	48.7
Total	100.0	100.0

(Chi-square = 23.6, d.f. = 2, $p = 0.0$.)

Table 4.5.10: Perception of Internet – "Internet Skills are Important for Getting a Good Job"

	Users (n=434) Percent	Non-Users (n=515) Percent
Disagree	11.1	16.3
Neutral	27.2	45.0
Agree	61.8	38.6
Total	100.0	100.0

(Chi-square = 9.1, d.f. = 2, p = 0.0.)

Additional analysis showed no significant relationship between experience and perception. Perception did not improve or deteriorate as people had more years of Internet experience. Also, the relationship between amount of use per week and perception was extremely weak ($r = 0.1$, $p = 0.0$).

4.5.1 *Internet and Politics*

So far the results showed that users thought that the Internet was positive to them and to the general society as a whole. Although there are people who try to avoid politics, it is an inherent part of society. Numerous scholars have speculated that the Internet can have a liberating effect. It enables swift mobilisation and rapid information dissemination. It tears down controls and facilitates political discourse. Will this happen?

In this light, respondents were asked how they thought the use of Internet might enable them in influencing politics. Overall, the results showed no difference between users and non-users — both groups did not believe that the Internet would provide political empowerment.

The proportion of users and non-users who thought that with the use of Internet, people could have more political power were 13.6% and 14.4%, respectively. Also, only 18.7% users and 14.8% non-users believed the Internet would give people more say in governance. The percentages climbed when they were asked if public officials would care more (users at 31.1% and non-users at 24.7%).

However, 40.3% users and 24.5% non-users felt that the Internet could help people better understand politics. It is important to note that a third or more of the respondents could not quite determine whether the Internet would have such an effect and chose a neutral stance in these questions. Details can be found in Table 4.5.11 through Table 4.5.14.

Table 4.5.11: Internet and Politics – "People Can Have More Political Power"

	Users (n=464) Percent	Non-Users (n=515) Percent
Disagree	51.8	46.4
Neutral	34.6	39.2
Agree	13.6	14.4
Total	100.0	100.0

(Chi-square test showed no significant difference between users and non-users.)

Table 4.5.12: Internet and Politics – "People Will Have More Say about What the Government Does"

	Users (n=434) Percent	Non-Users (n=515) Percent
Disagree	41.5	42.5
Neutral	39.9	42.7
Agree	18.7	14.8
Total	100.0	100.0

(Chi-square test showed no significant difference between users and non-users.)

Table 4.5.13: Internet and Politics – "People Can Better Understand Politics"

	Users (n=434) Percent	Non-Users (n=515) Percent
Disagree	21.9	28.9
Neutral	37.8	46.6
Agree	40.3	24.5
Total	100.0	100.0

(Chi-square = 27.5, d.f. = 2, p = 0.0.)

Table 4.5.14: Internet and Politics – "Public Officials Will Care More About What People Think"

	Users (n=434) Percent	Non-Users (n=515) Percent
Disagree	28.8	31.1
Neutral	40.1	44.3
Agree	31.1	24.7
Total	100.0	100.0

(Chi-square test showed no significant difference between users and non-users.)

4.5.2 Implications for IT Promoters and Policy Makers

Which comes first — usage or perception — is a chicken-and-egg question that cannot be answered with this set of data. It is reasonable to believe that they influence each other back and forth. For those interested in promoting the Internet, where usage is the bottom-line concern, this study confirms that changing perception is a possible way to increase usage. This, of course, is nothing new. However, the findings have implications on promotional efforts.

First, educational and promotional efforts should point out how the Internet can benefit users. It should be specific and pertain to individuals. Non-users know that the Internet is good for society, but they are not logging on. The perception gap between users and non-users is three times larger in their assessment of its personal utility as compared with their assessment of the more abstract societal utility. Non-users do not see the Internet as advantageous to their personal lives at this point. Therefore, the answer is in making things relevant to non-users.

Second, educational and promotional efforts should bear in mind the top three reasons for not using the Internet — no time, do not know how and not interested (mentioned in the Usage Pattern section). Of these three factors, perhaps the lack of know-how should be the focal point. This is because once a person becomes a user, barriers such as no time and not interested may just fade away. As this study shows, users are more likely to think that the Internet makes life more convenient (if it also means saving time) and is interesting.

In the meantime, more and more Singaporeans, both adults and children, will be exposed (even forced) to the use of the Internet in either the work or school environment. The use of the Internet will increasingly be the choice of the individual, and become part of daily life. There are good reasons to believe there will be lesser barriers to the utilisation of the new media either for personal or for societal purposes. It will also be interesting to observe whether the current positive attitude and perception of the Internet will last or if it is just a novelty.

At this point, Singaporeans, especially users, seem to view the Internet as a commercial, educational and entertainment tool. However, they do not regard the Internet as useful in influencing politics. This can be attributed to two things. First, this reflects the political culture in Singapore, characterised by a strong, paternalistic government and a de-politicised population. In this sense, it will be interesting as we compare the Singapore findings with those in other countries. Second, Singaporeans have not learned to exploit the characteristics of the Internet for political purposes. Only time will tell if the Internet will become a major political playing field as it has become in other countries.

4.6 Internet Regulation

4.6.1 Concerns over Internet Content

A large proportion of the respondents in the survey expressed a concern over undesirable content on the Internet. More than half of the

Internet users said that they were concerned about uncensored pornographic (64.1%), political (51.8%), religious (52.1%) and racial (57.1%) content on the Internet. Among non-users, the percentage was lower but still high, expressing concerns about uncensored content of pornographic (52.4%), political (42.1%), religious (41%) and racial (40%) nature (see Table 4.6.1 – 4.6.4).

While more users than non-users expressed a concern about undesirable content on the Internet, there were also more of them who thought otherwise. In other words, the users as a group were far more ready to express their views, either agreeing or disagreeing, than were non-users. This is not surprising since the former has a direct and personal experience with this new medium, while a substantial percentage of non-users took a "neutral" stand as they probably had no knowledge or were unsure of the situation.

Table 4.6.1: Concern about Uncensored Pornographic Content

	Users (n=434) Percent	Non-Users (n=515) Percent
Disagree	25.8	19.2
Neutral	10.1	28.3
Agree	64.1	52.4
Total	100.0	100.0

(Chi-square = 49.1, d.f. = 2, p = 0.0.)

Table 4.6.2: Concern about Uncensored Political Content

	Users (n=434) Percent	Non-Users (n=515) Percent
Disagree	26.0	17.5
Neutral	22.1	40.4
Agree	51.8	42.1
Total	100.0	100.0

(Chi-square = 37.37, d.f. = 2, p = 0.00.)

Table 4.6.3: Concern about Uncensored Religious Content

	Users (n=434) Percent	Non-Users (n=515) Percent
Disagree	22.6	13.6
Neutral	25.3	45.4
Agree	52.1	41.0
Total	100.0	100.0

(Chi-square = 43.28, d.f. = 2, p = 0.00.)

Table 4.6.4: Concern about Racial Content

	Users (n=434) Percent	Non-Users (n=515) Percent
Disagree	19.1	12.2
Neutral	23.7	47.8
Agree	57.1	40.0
Total	100.0	100.0

Chi-square test showed a significant difference between users and non-users. (Chi-square = 58.7, d.f. = 2, p = 0.0.)

4.6.2 *Who Should be Responsible?*

Respondents were asked who should be responsible for either reducing or avoiding undesirable Internet content. They were allowed to select more than one of the five choices – the individual, the family, the Internet Service Provider (ISP), the government and a central body comprising representatives of the public, government and ISPs. Chi-square test showed no difference between users and non-users in their choices. As shown in Table 4.6.5, a very large majority (92.2%) of respondents indicated that the individual should be responsible. This was followed by the Internet Service Provider (76.1%), a central body (66.3%), family (65%) and, finally, the government (63.8%).

In other words, in each case, at least a majority of close to 65% wanted some form of regulatory responsibility for that group. Collectively they seem to suggest a joint responsibility including all — the individual, the ISP, the central body, the family and the government. The heavy emphasis on the individual is especially noteworthy as it is contrary to what many commentators have said in the past — that Singaporeans have become overly dependent on the government and are not ready or mature enough to shoulder personal responsibility in such matters.

Table 4.6.5: Responsibility for Internet Regulation

	Yes		No	
	Freq	Percent	Freq	Percent
Individuals	875	92.2	74	7.8
Family	617	65.0	332	35.0
ISP	722	76.1	227	23.9
Government	605	63.8	344	36.2
Central Body	629	66.3	320	33.7

4.6.3 Rating System

On the question of whether there should be a rating system by content providers to advise the public so that they can make their own choices, more than a two-third majority (69.3%) agreed that there should be such a system (see Table 4.6.6). It is interesting to note that the majority in every age group in the survey shared this feeling (see Table 4.6.7). Also noteworthy is the finding that more of the higher educated tended to agree to the need for a rating system than did those with less education (see Table 4.6.8). The latter also showed a higher percentage that did not express a view on this issue.

Table 4.6.6: Need for Rating System by Content Providers

	Percent (n=949)
Disagree	7.8
Neutral	22.9
Agree	69.3
Total	100.0

Table 4.6.7: "Need for Rating System by Content Providers", by Age

Age Category	Disagree Percent	Neutral Percent	Agree Percent	Total Percent
18-24	12.4	25.7	61.8	100.0
25-34	4.8	17.2	78.0	100.0
35-44	2.4	15.6	82.0	100.0
45-54	8.4	30.1	61.4	100.0
55-64	13.7	30.1	56.2	100.0
65-74	12.2	31.7	56.1	100.0
75 and above	0	14.3	85.7	100.0

Table 4.6.8: "Need Rating System by Content Providers", by Education

	Disagree Percent	Neutral Percent	Agree Percent	Total Percent
Primary 6 or below	9.0	31.7	59.3	100.0
'O' level	8.6	23.1	68.3	100.0
A level	9.4	23.4	67.2	100.0
Diploma	4.1	16.6	79.3	100.0
Degree and above	6.0	12.9	81.0	100.0

4.6.4 Filter Options

As to whether the ISPs should offer some options for filtering out undesirable content, a two-third majority (66.3%) agreed that they should (see Table 4.6.9). Significantly, the young tend to be more supportive of the filter options than do their elders. As presented in Table 4.6.10, more of those aged between 18 and 44 (ranging from 65.1% to 74.5%) supported the filter options than did those aged between 45 and 74 (47.9% to 56.1%). Similarly, the higher educated were also more supportive of the filter options. Many more among those with a university degree or higher education (80.2%) agreed, compared with only 54.8% of those with Primary 6 or lower education (see Table 4.6.11). Here it is important to note that the younger people are not more averse to filters and that the higher educated probably see the benefits of these options more clearly and, therefore, favour them.

Table 4.6.9: Provision of Filter Options by ISPs

	Percent (n=949)
Disagree	9.5
Neutral	24.2
Agree	66.3
Total	100.0

Table 4.6.10: "Provision of Filter Options by ISPs", by Age

Age Category	Disagree Percent	Neutral Percent	Agree Percent	Total Percent
18-24	11.6	23.2	65.1	100.0
25-34	6.2	17.7	76.1	100.0
35-44	3.3	22.2	74.5	100.0
45-54	15.1	30.1	54.8	100.0
55-64	16.4	35.6	47.9	100.0
65-74	12.2	31.7	56.1	100.0
75 and above	0	14.3	85.7	100.0

Table 4.6.11: "Provision of Filter Options by ISPs", by Education

	Disagree Percent	Neutral Percent	Agree Percent	Total Percent
Primary 6 or below	9.5	35.7	54.8	100.0
'O' level	12.4	22.1	65.5	100.0
A level	7.0	26.6	66.4	100.0
Diploma	6.9	18.6	74.5	100.0
Degree and above	6.9	12.9	80.2	100.0

4.6.5 Implications for Government, ISPs, Content Providers and the Public

The study shows that a large proportion of respondents, both users and non-users, were concerned about undesirable content on the Internet. There are significant implications for policy consideration. It is pertinent to note that roughly two-thirds of the respondents wish to involve the government, the ISPs and a central body (comprising representatives of the public, the ISPs and the government) in regulating the Internet. In other words, they seem to seek a collective and multiple effort involving all three regulatory groups. This is in addition to self-regulation by the individual and the family, with a high rate of 9 out of 10 mentioning that the individual should be responsible.

What is also significant but not explored in the survey is the assumption that some authoritative groups can indeed regulate the Internet. Given the reality that these authoritative groups can hardly control the influx of content on the Internet in a highly open telecommunication environment such as Singapore's, this assumption needs deeper examination.

4.7 Importance of Information Sources

With the advent of the Internet, traditional media were forced to adapt and change. One nagging concern was whether the new medium would replace the traditional mass media, as well as interpersonal sources, such as friends, colleagues and family members.

4.7.1 Mass Media

Since the media could contribute in many ways, this study focused only on information provision. Respondents were asked how important the Internet, television, radio, newspapers and magazines were as sources of information.

The results indicated that there was no immediate concern that the Internet would replace traditional mass media. This was because Internet users seemed to be users of other kinds of media as well. For all the media listed, the proportions of users indicating a particular medium was important were either similar or higher than that indicated by non-users. This suggested that Internet users, being of higher socio-economic status (SES) in general, sought out information through various channels — more so than non-Internet users. Relevant statistics are presented in Tables 4.7.1 – 4.7.5:

Internet. A total of 71.2% users said that Internet was an important source of information. Meanwhile, 15.7% users said it was not important (see Table 4.7.1).

Television. Almost equal proportions of Internet users (73.7%) and non-users (74%) said that television was an important source of information (see Table 4.7.2).

Radio. As for radio, 69.4% and 64.7% of the Internet users and non-users said that it was an important source of information, respectively (see Table 4.7.3).

Newspapers. Newspapers were an important source of information to 86.6% and 78.8% of the Internet users and non-users, respectively (see Table 4.7.4).

Magazines. Compared to the other media, a lower percentage of respondents cited magazines as an important information source. Still, about 53% of the users and 45% of the non-users regarded it as the main important source (see Table 4.7.5).

Table 4.7.1: Internet as a Source of Information

	Users (n=434) Percent	Non-Users (n=515) Percent
Unimportant	15.7	45.6
Neutral	13.1	31.7
Important	71.2	22.7
Total	100.0	100.0

(Chi-square = 47.5, d.f. = 2, p = 0.0.)

Table 4.7.2: Television as a Source of Information

	Users (n=434) Percent	Non-Users (n=515) Percent
Unimportant	12.9	11.7
Neutral	13.4	14.4
Important	73.7	74.0
Total	100.0	100.0

(Chi-square test indicated no significant difference between users and non-users.)

Table 4.7.3: Radio as a Source of Information

	Users (n=434) Percent	Non-Users (n=515) Percent
Unimportant	12.0	14.0
Neutral	18.7	21.4
Important	69.4	64.7
Total	100.0	100.0

(Chi-square test indicated no significant difference between users and non-users.)

Table 4.7.4: Newspapers as a Source of Information

	Users (n=434) Percent	Non-Users (n=515) Percent
Unimportant	3.5	9.3
Neutral	9.9	11.8
Important	86.6	78.8
Total	100.0	100.0

(Chi-square test indicated no significant difference between users and non-users.)

Table 4.7.5: Magazines as a Source of Information

	Users (n=434) Percent	Non-Users (n=515) Percent
Unimportant	14.1	29.1
Neutral	32.5	25.8
Important	53.5	45.0
Total	100.0	100.0

(Chi-square = 10.93, d.f. = 2, p = 0.00.)

Table 4.7.6 ranks the importance of various media as information sources on a 5-point scale (with 1 being not important at all, and 5 being extremely important). Newspapers, which have been in existence for a few centuries, were ranked the top — scoring 4.1 by users and 3.8 by non-users. Both groups considered television the second most important source (3.8 for users and 3.7 for non-users). The Internet is ranked third by users and, of course, last by non-users. Even then, radio is not all that far behind (3.6 for users and 3.5 by non-users).

Table 4.7.6: Importance of Various Media as Information Sources

	Users (n=434) Mean	Non-Users (n=515) Mean
Newspaper	4.1	3.8
Television	3.8	3.7
Internet	3.7	2.7
Radio	3.6	3.5
Magazines	3.4	3.1

4.7.2 Interpersonal Sources

Mass media do not have to be the main source of information. Sometimes, people do receive information from interpersonal channels. In the above section, more Internet users, as compared with non-users, were found to rank mass media as important sources of information. However, the reverse seemed to be true when it came to interpersonal sources. Table 4.7.8 shows that 72.2% of the non-Internet users said friends were important sources of information (as compared to 65.2% of the Internet users). In Table 4.7.7, the percentages were very close — with 51.1% non-users and 52.5% of users stating friends as important information sources. Finally, in Table 4.7.9, 79.8% of non-users and 66.40% of users claimed family members and relatives as important sources.

Table 4.7.7: Colleagues as a Source of Information

	Users (n=434) Percent	Non-Users (n=515) Percent
Unimportant	12.9	22.5
Neutral	34.6	26.4
Important	52.5	51.1
Total	100.0	100.0

(Chi-square test indicated no significant difference between users and non-users.)

Table 4.7.8: Friends as a Source of Information

	Users (n=434) Percent	Non-Users (n=515) Percent
Unimportant	7.4	7.2
Neutral	27.4	20.6
Important	65.2	72.2
Total	100.0	100.0

(Chi-square test indicated no significant difference between users and non-users.)

Table 4.7.9: Family and Relatives as a Source of Information

	Users (n=434) Percent	Non-Users (n=515) Percent
Unimportant	7.6	5.8
Neutral	26.0	14.4

Table 4.7.9: Family and Relatives as a Source of Information (cont'd)

	Users (n=434) Percent	Non-Users (n=515) Percent
Important	66.4	79.8
Total	100.0	100.0

(Chi-square = 7.2, d.f. = 2, p = 0.0.)

4.7.3 Implications for Traditional Mass Media and Interpersonal Communication

The findings suggest that traditional sources of information, such as newspapers and television, are in no danger of being replaced by the Internet at the time of the survey. Even among Internet users, this new medium is only ranked third. People, especially those better informed, do not seek out information from one place only — they rely on multiple sources. It does not seem that one medium will become the champion and eliminate the rest in this case. The concern, however, is that the time spent on each may subsequently be reduced — after all, a person has only 24 hours a day. This question cannot be answered at this point. However, this study has recorded the amount of time people spent on various activities in 1999. Longitudinal data will be able to show if the shares have shrunk over subsequent years.

Another prominent observation is that Internet-users seem to prefer mass media and colleagues as sources of information, while non-users rely on personal contacts like friends, family and relatives — what sociologists call "primary groups". (In fact, family and relatives were indicated as the most important sources among non-users). Media and workplace environment is less concordant, and ideas and opinions from the media and colleagues are mixed and even divergent — some are supportive and others may not share the same attitudes and beliefs. Friends, family and relatives, on the other hand, form a relatively supportive and harmonious environment. This study shows that the communication environment differs to some extent between Internet users and non-users. The former — being younger, single and of higher socio-economic status in terms of education, income, and housing — are more open and receptive to a less concordant communication environment. They tend to be working people or students. They are required to pay attention to various sources and deal with information from diverse sources to form their own opinion or to do their jobs. Many of them may be opinion leaders in the multiple-step flow of information process. On the other hand, non-users tend to be older, married and of lower socio-economic status. Their social network tends to be close-knit and "primary" in nature. They are more likely to share information with other members of their primary group. Their information needs are likely to be of a personal nature and not job-based.

4.8. Trust of Institutions

The final section of this report addresses the issue of trust. Would the Internet boost or erode people's confidence in various social institutions, such as the media, the government, business corporations and religious organisations?

In this case, respondents were asked if they trusted the news obtained from television, radio, newspapers and the Internet. A high proportion of respondents — both Internet users and non-users — replied that they trusted the news from these media.

Consistently, more than 74% of the respondents said that they trusted news from television, radio and newspapers. They were more suspicious of online news — still, 60% users and 46.8% non-users said they trusted it.

Table 4.8.1: Trust in Television News

	Users (n=434) Percent	Non-Users (n=515) Percent
Do not trust	10.8	7.0
Neutral	12.4	16.3
Trust	76.7	76.7
Total	100.0	100.0

(Chi-square = 9.4, d.f. = 2, p = 0.0.)

Table 4.8.2: Trust in Radio News

	Users (n=434) Percent	Non-Users (n=515) Percent
Do not trust	10.6	7.4
Neutral	13.8	18.4
Trust	75.6	74.2
Total	100.0	100.0

(Chi-square test indicated no significant difference between users and non-users.)

Table 4.8.3: Trust in Newspapers' News

	Users (n=434) Percent	Non-Users (n=515) Percent
Do not trust	7.8	6.4
Neutral	13.6	16.9
Trust	78.6	76.7
Total	100.0	100.0

(Chi-square test indicated no significant difference between users and non-users.)

Table 4.8.4: Trust in Online News

	Users (n=434) Percent	Non-Users (n=515) Percent
Do not trust	9.5	14.0
Neutral	30.7	39.2
Trust	59.9	46.8
Total	100.0	100.0

(Chi-square test indicated no significant difference between users and non-users.)

In addition to the media, respondents indicated if they trusted the government, business corporations and religious organisations. For all three institutions, the percentages of people who said they did not trust them were relatively small — ranging from 6% to 18.1%. A total of 71.2% users said they trusted the government. Non-users lagged behind, but still 70.3% of the non-users concurred. The business community, however, did not fare as well as the government — 45.4% users and 39% of the non-users expressed trust in businesses and corporations. Similarly, 41% users and 41.6% of the non-users trusted religious organisations. It is noteworthy that more than 40% of the respondents were either neutral or could not decide if they could trust businesses and religious organisations. (This is not surprising as business and religious organisations, respectively, do not form coherent categories. For instance, one is likely to find that some religious organisations can be trusted, but not others).

Table 4.8.5: Trust in Government

	Users (n=434) Percent	Non-Users (n=515) Percent
Do not trust	6.9	6.0
Neutral	21.9	23.7
Trust	71.2	70.3
Total	100.0	100.0

(Chi-square test indicated no significant difference between users and non-users.)

Table 4.8.6: Trust in Business Companies/Corporations

	Users (n=434) Percent	Non-Users (n=515) Percent
Do not trust	12.2	17.7
Neutral	42.4	43.3
Trust	45.4	39.0
Total	100.0	100.0

(Chi-square test indicated no significant difference between users and non-users.)

Table 4.8.7: Trust in Religious Organisations

	Users (n=434) Percent	Non-Users (n=515) Percent
Do not trust	16.8	18.1
Neutral	42.2	40.4
Trust	41.0	41.6
Total	100.0	100.0

(Chi-square test indicated no significant difference between users and non-users for both adult and student samples.)

4.8.1 Implications for Institutions

The findings in this section further affirm that traditional media are still important to people. Users and non-users alike still believed that news from traditional media can be trusted. In fact, news from newspaper tops the chart, followed, in order, by television, radio and the Internet. Three-quarters or more of users trusted news from traditional media. Online news lags behind by 15%. As a new medium, people are not sure if online information is reliable. It is easy to know who has control and influence over news from traditional media in the Singapore context. However, this is not the same with online news — all kinds of people can disseminate online news or even tamper with online news. We will have to continue to monitor if trustworthiness of online news will improve in the future.

Some observers speculate that the wider spread of the Internet may politicise the population and disrupt social stability. The findings in this survey do not show alarming signs. Internet users showed a low level of alienation. They did not find that the use of the Internet had led to political empowerment. Generally, only six or seven people in a hundred said that they did not trust the government. Fewer than one in five said that they distrusted business corporations or religious organisations. All signs seem to indicate Singaporeans are generally pro-establishment. Given the fact that users and non-users do not differ in these aspects, there is no indication that the use of the Internet is associated with greater challenge to the establishment in Singapore at present.

CHAPTER 5

Findings (Student Survey)

5.1 Sample Demographics

5.1.1 Population and Sampling

For the 1999 survey, permission was obtained from the Ministry of Education (MOE) and respective school Principals to conduct surveys in secondary schools. Out of the 152 secondary schools, 33 were selected at random and 19 responded positively before the closing date. Two classes from each school took part in the survey. A total of 1,330 Secondary One school students filled out the questionnaires.

5.1.2 Demographic Characteristics of Respondents

Although 1,330 Secondary One students were surveyed, analysis was conducted only on 1,251 valid and completed questionnaires. Table 5.1.1 through Table 5.1.3 compare the student sample (Secondary One) to the general student population (Secondary One to Three) in terms of gender, race and type of schools. (Population figures on Secondary One to Three were obtained from MOE as at September 2000.)

Table 5.1.1: Gender Distribution – Student Sample vs. Population

Gender	Sample (%)	Population (%)
Male	43.0	52.0
Female	57.0	48.0
Total	100.0	100.0

Table 5.1.2: Race – Sample vs. Population

Race	Sample (%)	Population (%)
Chinese	71.5	72.7
Malay	15.9	17.7
Indian	10.6	8.1
Others	2.0	1.6
Total	100.0	100.0

Table 5.1.3: Stream – Sample vs. Population

Type of Stream	Sample (%)	Population (%)
Special	6.2	9.7
Express	58.6	51.3
Normal	35.2	39.1
Total	100.0	100.0

5.2. Internet Usage Patterns

5.2.1 Internet Use

The study found a high number of computer and Internet users among Secondary One students (see Tables 5.2.1 and 5.2.2). Close to 93% of the students were computer users and 71.3% were Internet users. On average, users had 1.6 years of Internet experience.

Table 5.2.1: Proportion of Computer Users and Non-Users

	Percent (n=1250)
Yes	92.9
No	7.1
Total	100.0

Table 5.2.2: Proportion of Internet Users and Non-Users

	Percent (n=1251)
Yes	71.3
No	28.7
Total	100.0

The computer and Internet penetration rates among Secondary One students were higher than those found among the adult population, although they had less experience. The proportion of computer users among the students was close to 93%, while it was only 57% among the adults. Furthermore, the percentage of Internet users was much higher among Secondary One students (71.3%) than among adults (45.7%).

This reflects the success of governmental effort in incorporating IT in the school curriculum and in creating an Internet-savvy population. Given the encouragement by the government and the schools, we can expect to see more users among this group of survey participants over the years.

5.2.2 *Demographics of Users and Non-Users*

Overall, the data showed no gender divide among the students. However, students from families of better socio-economic status, in terms of income, housing and parents' education were much more likely to use the Internet than their counterparts.

Gender. Chi-square test indicated no significant difference between male and female students in Internet adoption. About three-quarters of both male and female students were users (see Table 5.2.3). Whereas in the adult survey, a gender gap was found in that significantly more males than females were Internet users.

Table 5.2.3: Internet Users/Non-Users, by Gender

	Users Percent	Non-Users Percent	Total (n = 1251) Freq	Percent
Male	73.2	26.8	538	100.0
Female	69.8	30.2	713	100.0

(Chi-square test indicated no significant difference.)

Race. The study found that Malay students trailed behind the other races by over 10% in Internet use. Among the Chinese, Indian and "Other" racial groups, 72–74% of the students were Internet users; whereas for the Malays, only 60% were Internet users (See Table 5.2.4). In order to have a better understanding of the ethnic differences in Internet use, we further analysed Internet use at home and at school among the students.

The statistics showed that about 36% of the Malay student respondents were Internet users, compared with 40% among Indians, and over 45% among Chinese. The difference, however, was not statistically significant (see Table 5.2.5). This is not the case when we come to the use of the Internet at home, where the differences among various ethnic groups were found to be statistically significant (see Table 5.2.6). The percentage of students using the Internet at home among the Chinese, Indian and Other racial groups were 55%, 52% and 60%, respectively. For the Malays, only 34% of the students used the Internet at home.

The above statistics suggest that it is mainly within the home environment that Malay students are disadvantaged in Internet access. This may be due to differences in cultural or social capital provided by the family to the student at home. As the education system is promoting the Internet to all students, the Internet playing field is more or less level in school. Yet, the digital divide along racial lines remains. There are reasons to believe that when the spread of Internet use in school reaches the level of 90% or higher, the digital divide along the ethnic lines may become somewhat moderated.

Table 5.2.4: Internet Users/Non-Users, by Race

	Users Percent	Non-Users Percent	Total (n = 1251)	
			Freq	Percent
Chinese	73.4	26.6	895	100.0
Malay	59.8	40.2	199	100.0
Indian	74.2	25.8	132	100.0
Other	72.0	28.0	25	100.0

(Chi-square = 15.4, d.f. = 3, p = 0.0.)

Table 5.2.5: Internet Users/Non-users, by Race at School

	Users Percent	Non-Users Percent	Total (n = 1251)	
			Freq	Percent
Chinese	45.7	54.3	895	100.0
Malay	36.2	63.8	199	100.0
Indian	40.2	59.8	132	100.0
Others	44.0	56.0	25	100.0

(Chi-square test indicated no significant difference between users and non-users.)

Table 5.2.6: Internet Users/Non-users, by Race at Home

	Users Percent	Non-Users Percent	Total (n = 1251)	
			Freq	Percent
Chinese	55.3	44.7	895	100.0
Malay	34.2	65.8	199	100.0
Indian	51.5	48.5	132	100.0
Others	60.0	40.0	25	100.0

(Chi-square = 29.8, d.f. = 3, p = 0.0.)

Income. [9] The findings indicated that students from richer families were more likely to be Internet users. As shown in Table 5.2.7, the proportions of users tended to be smaller in lower income brackets as compared to the upper income levels. For the $2,000-or-less income bracket, the percentages of users and non-users were 56.3% and 43.7%, respectively. At the $2,000-$3,000 level, there were 64.8% users and 35.2% non-users. We saw about 80% users and 20% non-users at income levels beyond $3,000.

Table 5.2.7: Internet Users/Non-Users, by Monthly Family Income

	Users Percent	Non-Users Percent	Total (n = 683)	
			Freq	Percent
$2000 or less	56.3	43.7	183	100.0
$2001 - $3000	64.8	35.2	179	100.0
$3001 - $4000	82.7	17.3	110	100.0
$4001 - $5000	86.2	13.8	58	100.0
$5001 - $6000	79.4	20.6	34	100.0
$6001 - $7000	81.5	18.5	27	100.0
Above $7000	92.4	7.6	92	100.0

(Chi-square = 60.6, D.F. = 6, p = 0.0.)

Housing. The observation that Internet users were more affluent than non-users was further confirmed by the type of housing they lived in. Table 5.2.8 presents the distribution of housing type of students. The percentage of users rose from 43.5% in HDB 1- or 2-room flats to 58.8% in HDB 3-room flats. The figure further increased steadily in better housing types — 67.4%, 74.9% and 89.5% in HDB 4-room, HDB 5-room/HDB executive/HUDC and private apartments, respectively.

---

9 Respondents who indicated that they did not know the amount and those who refused to answer were omitted from the analysis.

Table 5.2.8: Internet Users/Non-Users, by Housing Type

	Users Percent	Non-Users Percent	Total (n = 1251)	
			Freq	Percent
HDB 1- or 2-Room	43.5	56.5	23	100.0
HDB 3-Room	58.8	41.2	221	100.0
HDB 4-Room	67.4	32.6	442	100.0
HDB 5-Room/ HDB Executive/ HUDC	74.9	25.1	354	100.0
Private	89.5	10.5	209	100.0
Others	100.0	0	2	100.0

(Chi-square = 65.5, d.f. = 5, p = 0.0.)

Education. Students' Internet usage was related to the educational level of their parents (see Tables 5.2.9 and 5.2.10). For students whose parents had primary or below education, only 58% were users. Among those whose fathers had O-level or A-level education, about 70% were users. Similarly, among students whose mothers had attained O-level or A-level education, we found 73% and 85% users, respectively. Finally, among students whose parents had completed tertiary education, the percentages of users were above 80%.

Table 5.2.9: Internet Users/Non-Users, by Father's Education

	Users Percent	Non-Users Percent	Total (n = 770)	
			Freq	Percent
Primary or below	58.5	41.5	176	100.0
'O' level or equivalent	69.8	30.2	301	100.0
A level	70.7	29.3	99	100.0
Diploma	91.3	8.7	92	100.0
Degree and above	87.3	12.7	102	100.0

(Chi-square = 45.7, d.f. = 4, p = 0.0.)

Table 5.2.10: Internet Users/Non-Users, by Mother's Education

	Users Percent	Non-Users Percent	Total (n = 815) Freq	Percent
Primary or below	57.7	42.3	239	100.0
'O' level or equivalent	73.0	27.0	344	100.0
A level	84.8	15.2	105	100.0
Diploma	84.7	15.3	59	100.0
Degree and above	80.9	19.1	68	100.0

(Chi-square = 39.7, d.f. = 4, p = 0.0.)

Age. The student sample had a narrow age range: most of them were between 12- and 15-years-old. Comparisons between age groups would not be meaningful and are, therefore, not reported here.

In sum, while the 1999 adult survey showed a gender gap, this was not found among Secondary One students. Similar to the adult study, however, the student survey also showed some evidence of the "digital divide" along the social-economic status (SES) line, as measured by family income, housing type and parents' education. The data seemed to indicate that the socio-economic disparities were behind the ethnic differences in Internet usage.

5.2.3 Internet Skills

The respondents were asked if they could perform the following activities: sending e-mail, print and save Internet files, use search engines, download information (including graphics and software programmes), participate in online chat groups and set up a web page. A total of 87.4% users were able to perform three or more skills.

The findings show that most users knew how to send e-mail (98.6%), print and save files (95.9%), and use search engines (90.3%). A high proportion could also join online chat groups (71.7%) and download information (81.3%). Compared to other Internet skills, setting up a web site was not as widely known — only 42.4% of the students were able to do it (see Table 5.2.11).

Table 5.2.11: Internet Skills among Users

	Percent (n=892)		Percent (n=892)
E-mail		Online chat	
Yes	98.6	Yes	71.7
No	1.4	No	28.3
Print/save		Download[a]	
Yes	95.9	Yes	81.3
No	4.1	No	18.7
Search engines		Set up web page	
Yes	90.3	Yes	42.4
No	9.7	No	57.6

a. Include downloading of computer programmes, graphics, etc.

5.2.4 Internet Activities

This section describes where respondents typically logged on to the Internet (see Table 5.2.12) and how much time they spent on Internet activities (see Table 5.2.13). As shown in Table 5.2.13, users spent an average of 5.4 hours per week on the Internet. Home use was prominent among Secondary One students, as on the average, they logged on 3.3 hours per week from home. They also reported using the Internet from other places, such as school (1 hour per week) and library (0.5 hour per week). This points toward the importance of the family and parents in monitoring young people's Internet access.

Secondary One students' weekly Internet usage was about half of that of the adults. The adults had a much higher usage mainly because of work — half of their Internet time was chalked up at the workplace.

Table 5.2.12: Internet Access at Various Locations

	Yes Percent	No Percent	Total (n=892) Percent
Home	72.4	27.6	100.0
School	61.1	38.9	100.0
Library	34.0	66.0	100.0
Others	21.9	78.1	100.0

Table 5.2.13: Time Use at Various Locations

	Average Hours Per Week
Home	3.3
School	1.0
Library	0.5
Others	0.5
Total	5.4

Student users spent most of their Internet time on information search for personal reasons (1.4 hours per week) and online chats (1.3 hours per week). Entertainment and information search for school work each took up about one hour per week. The other activities took up less than an hour per week (see Table 5.2.14).

Table 5.2.14: Time Spent on Various Internet Activities

Internet Activities	Hours Per Week
E-mail	0.8
Online discussion/chat	1.3
Entertainment	1.1
Info seek — work/school	1.1
Info seek – personal	1.4
Transaction	0.1
Online news	0.3
Total	6.1

5.3 Lifestyle and Well-Being

One of the objectives of this study is to monitor how the Internet could affect lifestyle (such as daily activities, interaction with people) and well-being. With one survey, it would be difficult to make conclusive statements about the Internet's impact. The 1999 survey would, however, provide baseline measurements and allow comparisons with findings from subsequent years.

5.3.1 Activities

Table 5.3.1 lists the amount of time respondents spent on 12 different activities. Statistical tests showed no significant differences between student users and non-users in all 12 activities, although users reported spending slightly more time on each activity (except television viewing) than non-users. This pattern was different from that found among the adults, where users were likely to spend more time on various media-related activities, but less face-to-face interaction with family, compared with non-users.

Table 5.3.1: Time Spent on Various Activities (hours per week)

	Users	Non-Users
Books	5.4	4.8
Video/PC Games	3.7	3.2
Tapes/CDs/MDs/Records	4.2	3.9
Newspapers	1.9	1.6
Magazines	1.5	1.4
Radio	5.1	4.6
Telephone	3.8	3.6
Television	7.9	8.2
Cinema	1.8	1.7
Face-to-face: Friends	5.9	5.7
Face-to-face: Family	15.9	14.2
Exercise/Sports	4.5	4.2

T-tests showed no difference (at alpha = 0.05 level) between users and non-users in all activities.

5.3.2 Social and Family Life

Users were surveyed on their interaction with friends and family (see Table 5.3.2). The results revealed a good sign for friendship. A total of 55.3% of users said they had made more friends, as compared to 7.2% students who said they had fewer. Also, 36.2% of the users said they were spending more time with close friends, as opposed to 16.1% who said otherwise.

Close to 60% of users said that Internet use had not affected the amount of time they spent with their families. About 24% users said that the new medium had reduced their time spent with family, as compared with about 17% of student users who said they had experienced an increase.

Table 5.3.2: Reported Change in Social Interaction with Internet Use

	Percent (n=888)
Have more friends	
Less	7.2
No change	37.5
More	55.3
Time spent with close friends	
Less	16.1
No change	47.7
More	36.2
Time spent with family	
Less	23.6
No change	59.6
More	16.8

To further investigate the impact of the Internet on family life, the researchers compared the quality of family interaction between users and non-users. There were no significant differences between users and non-users in three of the categories. The proportions were similar between users and non-users who said that they and their families

(1) listened to each other (Table 5.3.4), (2) knew what each other were going through (Table 5.3.5), and (3) were patient with each other (Table 5.3.7) were similar. On two other indicators, users were found to be more likely than non-users to say that they and their families had time for each other (Table 5.3.3) and they shared ideas and opinions (Table 5.3.6). In each of these five situations, a large number of both users and non-users either were neutral or reported a positive experience in family communication. Also, the percentage of users who had negative experiences was slightly lower than that of the non-users.

Overall, this set of questions indicated that there was no statistical difference between users and non-users in the quality of family interaction. This was consistent with the findings from the adult survey.

Table 5.3.3: Family Interaction – "You and Family are Too Busy to Talk to Each Other"

	Users (n=855) Percent	Non-Users (n=355) Percent
Disagree	67.6	59.2
Neutral	19.2	23.7
Agree	13.2	17.2
Total	100.0	100.0

(Chi-square = 8.0. d.f. = 2, p = 0.0.)

Table 5.3.4: Family Interaction – "You and Family Listen to Each Other When There is Disagreement"

	Users (n=884) Percent	Non-Users (n=356) Percent
Disagree	23.1	23.3
Neutral	28.2	30.9
Agree	48.8	45.8
Total	100.0	100.0

(Chi-square test indicated no significant differences between users and non-users.)

Table 5.3.5: Family Interaction – "You and Family Do Not Really Understand What Each Other is Going Through"

	Users (n=883) Percent	Non-Users (n=353) Percent
Disagree	44.5	38.8
Neutral	32.4	34.6
Agree	23.1	26.6
Total	100.0	100.0

(Chi-square test indicated no significant differences between users and non-users.)

Table 5.3.6: Family Interaction – "You and Family Share Ideas and Opinions"

	Users (n=884) Percent	Non-Users (n=354) Percent
Disagree	12.6	17.5
Neutral	22.3	25.7
Agree	65.2	56.8
Total	100.0	100.0

(Chi-square = 8.5, d.f. = 2, $p = 0.0$.)

Table 5.3.7: Family Interaction – "You and Family are Patient with One Another"

	Users (n=885) Percent	Non-Users (n=356) Percent
Disagree	15.1	16.3
Neutral	29.9	31.2
Agree	54.9	52.5
Total	100.0	100.0

(Chi-square test indicated no significant differences between users and non-users.)

5.3.3 School Performance

As shown in Table 5.3.8, the Internet was generally perceived as enhancing productivity. About 60% of users said that Internet use had helped elevate their productivity level, while only 9.9% said it was detrimental to productivity. This finding was consistent with that from the adult survey.

Table 5.3.8: Reported Change in Productivity with Internet Use

	Percent (n=888)
Less	9.9
No Change	29.7
More	60.4
Total	100.0

In addition, a comparison was made between users and non-users in school performance, based on their reported exam results from the last semester (on 100% basis). As shown in Table 5.3.9, Internet users were found to have performed slightly better than non-users in examination results. While statistically significant, the difference was, however, rather small and should not be taken as conclusive.

Table 5.3.9: Internet Use and Examination Results

	Users	Non-Users
Mean	66.0	64.8

($t = 2.2$, d.f. = 1077, $p = 0.0$.)

5.3.4 Well-Being

Six items were used to measure alienation — if people felt helpless, if they thought they could influence the government, if they believed they could change things in the world, if they thought life was getting worse, if they could count on others and if others cared. Results are presented in Table 5.3.10 through Table 5.3.15. The differences between users and non-users were not statistically significant.

Again, this finding was consistent with that from the adult survey. At the time of the survey, the research team found no indication that the use of Internet was related to alienation. These measurements could serve as baselines to see if there may be changes among this group of secondary school students in the coming years.

Table 5.3.10: Alienation Measurement – "Feeling Helpless"

	Users (n=887) Percent	Non-Users (n=357) Percent
Disagree	27.3	26.3
Neutral	40.8	45.7
Agree	31.9	28.0
Total	100.0	100.0

(Chi-square test indicated no significant difference between users and non-users.)

Table 5.3.11: Alienation Measurement – "Average Citizen Can Have Influence on Government"

	Users (n=885) Percent	Non-Users (n=356) Percent
Disagree	10.2	8.4
Neutral	45.4	52.2
Agree	44.4	39.3
Total	100.0	100.0

(Chi-square test indicated no significant difference between users and non-users.)

Table 5.3.12: Alienation Measurement – "Can Change Course of World Events"

	Users (n=885) Percent	Non-Users (n=356) Percent
Disagree	7.1	6.5
Neutral	42.9	49.7
Agree	49.9	43.8
Total	100.0	100.0

(Chi-square test indicated no significant difference between users and non-users.)

Table 5.3.13: Alienation Measurement – "Average People are Getting Worse"

	Users (n=883) Percent	Non-Users (n=356) Percent
Disagree	14.5	13.2
Neutral	56.5	58.7
Agree	29.0	28.1
Total	100.0	100.0

(Chi-square test indicated no significant difference between users and non-users.)

Table 5.3.14: Alienation Measurement – "Don't Know Who to Count On"

	Users (n=883) Percent	Non-Users (n=356) Percent
Disagree	14.9	12.6
Neutral	40.6	45.1
Agree	44.5	42.3
Total	100.0	100.0

(Chi-square test indicated no significant difference between users and non-users.)

Table 5.3.15: Alienation Measurement – "Most People Don't Care What Happens to Others"

	Users (n=885) Percent	Non-Users (n=357) Percent
Disagree	18.5	16.0
Neutral	34.8	38.4
Agree	46.7	45.7
Total	100.0	100.0

(Chi-square test indicated no significant difference between users and non-users.)

5.4 Perceptions of the Internet

The perception of the Internet was included in the SIP survey because it could help explain usage patterns and social impact. It is reasonable to believe that a positive view of the Internet may lead to more usage and vice versa. Meanwhile, experience with the Internet may also affect people's perception of the medium. Hence, the respondents were asked a series of questions pertaining to their perception of the Internet.

The first five items were about how they viewed the Internet in relation to themselves — if they thought the Internet was important, useful, interesting, easy to use and convenient to them. Results are presented in Tables 5.4.1 to 5.4.5. As expected, users were more likely than non-users to see the Internet in a positive light: important (74.3%), useful (90.4%), interesting (83.8%), easy to use (82.8%) and makes life more convenient (73.4%). In contrast, non-users were less likely to agree — although the percentages were quite high: important (62.1%), useful (78.6%), interesting (72.9%), easy to use (56.5%) and making life more convenient (60.5%).

There was a difference of about 11 to 13 percentage points between users and non-users for all items — with the exception "easy to use," which had a 26.3% points gap. This seemed to suggest that non-users were relatively "ready" to embrace the Internet, but there may be a barrier to adopting the Internet because of the perceived difficulty of use.

It is interesting to note that many non-users also held a positive perception of the Internet. The contrast between users and non-users for these five items was not as enormous as that in the adult survey, which could have anywhere between 46 to 55 percentage point difference.

Table 5.4.1: Perception of Internet – "Internet is Unimportant"

	Users (n=883) Percent	Non-Users (n=359) Percent
Disagree	74.3	62.1
Neutral	15.6	25.4
Agree	10.1	12.5
Total	100.0	100.0

(Chi-square = 19.8, d.f. = 2, p = 0.0.)

Table 5.4.2: Perception of Internet – "Internet is a Useful Tool"

	Users (n=886) Percent	Non-Users (n=359) Percent
Disagree	3.2	4.2
Neutral	6.4	17.3
Agree	90.4	78.6
Total	100.0	100.0

(Chi-square = 36.3, d.f. = 2, p = 0.05.)

Table 5.4.3: Perception of Internet – "Internet is Uninteresting"

	Users (n=884) Percent	Non-Users (n=358) Percent
Disagree	83.8	72.9
Neutral	10.9	18.2
Agree	5.3	8.9
Total	100.0	100.0

(Chi-square = 19.5, d.f. = 2, p = 0.05.)

Table 5.4.4: Perception of Internet: "Internet is Easy to Use"

	Users (n=888) Percent	Non-Users (n=359) Percent
Disagree	4.1	10.4
Neutral	13.1	33.1
Agree	82.8	56.5
Total	100.0	100.0

(Chi-square = 94.6, d.f. = 2, $p = 0.05$.)

Table 5.4.5: Perception of Internet: "Internet Makes Life More Convenient"

	Users (n=888) Percent	Non-Users (n=359) Percent
Disagree	5.3	5.9
Neutral	21.3	33.7
Agree	73.4	60.5
Total	100.0	100.0

(Chi-square = 22.2, d.f. = 2, $p = 0.05$.)

In addition to the above five perception questions, respondents were asked if they thought the Internet was good for society, for staying informed, for entertainment purposes, for business activities and for getting a job. The data showed that the proportions of users who agreed with these statements were 62.7%, 85.9%, 83.7%, 66% and 54.7%, respectively. Meanwhile, the percentages for non-users who agreed were 60.2%, 80.8%, 75.8%, 75.7% and 60.4%, respectively. The differences between users and non-users for these items ranged from 2.5% to 9.7% points — relatively smaller than the first five perception questions. Details can be found in Tables 5.4.6 – 5.4.10.

Again, the results showed that both users and non-users were likely to think positively about the Internet. There was no statistically significant difference between the two groups with regards to the first item "Internet is a good thing for society" and the last item "Internet skills are important for getting a good job". Users were more likely to say that the Internet was useful for staying informed and for entertainment purposes. However, the gaps between the two groups were only 5.1% and 7.9% points, respectively. Rather unexpectedly, more non-users than users believed that the Internet was beneficial for business activities.

While the results in the 1999 adult survey also showed that users and non-users were rather close in their outlook with regard to these five questions, adult users were consistently more positive in general. The Secondary One student sample seemed not to conform to that pattern. For two items, users were more positive, and in one, it was non-users who were more positive.

Table 5.4.6: Perception of Internet – "Internet is a Good Thing for Society"

	Users (n=888) Percent	Non-Users (n=359) Percent
Disagree	6.1	8.1
Neutral	31.2	31.8
Agree	62.7	60.2
Total	100.0	100.0

(Chi-square test indicated no significant difference between users and non-users.)

Table 5.4.7: Perception of Internet – "Internet is Useful to Stay Informed"

	Users (n=888) Percent	Non-Users (n=359) Percent
Disagree	2.7	3.1
Neutral	11.4	16.2
Agree	85.9	80.8
Total	100.0	100.0

(Chi-square = 5.5, d.f. = 2, $p = 0.06$.)

Table 5.4.8: Perception of Internet – "The Internet is Great for Entertainment Purposes"

	Users (n=890) Percent	Non-Users (n=359) Percent
Disagree	2.8	6.7
Neutral	13.5	17.5
Agree	83.7	75.8
Total	100.0	100.0

(Chi-square = 14.7, d.f. = 2, p = 0.0.)

Table 5.4.9: Perception of Internet – "Internet Makes Business/Commercial Activities More Convenient"

	Users (n=888) Percent	Non-Users (n=358) Percent
Disagree	3.9	4.5
Neutral	30.1	19.8
Agree	66.0	75.7
Total	100.0	100.0

(Chi-square = 13.5, d.f. = 2, p = 0.0.)

Table 5.4.10: Perception of Internet – "Internet Skills are Important for Getting a Good Job"

	Users (n=886) Percent	Non-Users (n=359) Percent
Disagree	5.6	5.6
Neutral	39.6	34.0
Agree	54.7	60.4
Total	100.0	100.0

(Chi-square test indicated no significant difference between users and non-users.)

5.5 Importance of Information Sources

To most adult Singaporeans, the Internet is a relatively new medium. It was only in 1994 that the public gained access to this new medium. The population in general have relied on traditional media such as newspapers and television for information most of their lives. The experience of the younger generation of Singaporeans has been quite different, especially given the strong encouragement of Internet use in school with full support of the government. How would young Singaporeans rank the various media available to them? Would they rely on the Internet and ignore the other media?

5.5.1 Mass Media

Since the media is able to satisfy multiple needs (e.g. entertainment, escape), this study focused only on information provision. Students were asked how important were the Internet, television, radio, newspapers and magazines as sources of information.

With the top ranking, the Internet was selected by 78.1% of users as an important source of information. Newspapers came in second, with 69.7% users saying they were important. After newspapers, came television (66.3%), radio (50.8%) and, finally, magazines (49.4%). The order was different for non-users: newspapers (61.7%), television (54.5%), the Internet (53.5%), radio (43.2%) and magazines (39.9%). Details can be found in Tables 5.5.1 to 5.5.5.

Two patterns are of interest here. Firstly, we note that non-users as a group scored lower in every medium in comparison with users. This suggests that this group is generally not as interested in the media as are the users. We do not have a simple explanation for this association between the use of the Internet and the emphasis on various mass media as information sources. It is possible that those who show a strong need for information (i.e. information need) will source for needed information from the Internet as well as other sources. On the other hand, the association could be due to differences in the type of school one attends or one's family background.

The other finding is that the Internet was rated by the largest number of users as an important source, in comparison with conventional mass media. Even among non-users, the Internet received almost equal support as did television (53.5% compared with 54.5%) as an important source.

In comparison, results from the adult survey did not show that the traditional media were threatened by the Internet as a main source of information. In fact, adult users were likely to retrieve information from a variety of sources and the Internet was just one of them.

The scenario was quite different among the Secondary One respondents: more than three-quarters of users regarded the Internet as an important source, more than those choosing other media. However, it is early to conclude that the Internet will remain the most important source when this young generation grows up. After all, most of the respondents were 12 to 13-year-old children at the time of the survey. Understandably, they might not have the need or the cognitive ability to deal with certain media at this age. As emphasised earlier in this report, this longitudinal study will allow the researchers to follow up with these students and see if there is any change in their perception and use of various information sources as they grow older.

Table 5.5.1: Internet as a Source of Information

	Users (n=886) Percent	Non-Users (n=357) Percent
Unimportant	4.5	14.3
Neutral	17.4	32.2
Important	78.1	53.5
Total	100.0	100.0

(Chi-square = 80.7, d.f. = 2, p = 0.0.)

Table 5.5.2: Television as a Source of Information

	Users (n=886) Percent	Non-Users (n=358) Percent
Unimportant	7.6	11.2
Neutral	26.2	34.4
Important	66.3	54.5
Total	100.0	100.0

(Chi-square = 15.5, d.f. = 2, p = 0.0.)

Table 5.5.3: Radio as a Source of Information

	Users (n=886) Percent	Non-Users (n=359) Percent
Unimportant	14.9	19.2
Neutral	34.3	37.6
Important	50.8	43.2
Total	100.0	100.0

(Chi-square = 6.8, d.f. = 2, p = 0.0.)

Table 5.5.4: Newspapers as a Source of Information

	Users (n=885) Percent	Non-Users (n=358) Percent
Unimportant	6.7	12.0
Neutral	23.6	26.3
Important	69.7	61.7
Total	100.0	100.0

(Chi-square = 12, d.f. = 2, p = 0.0.)

Table 5.5.5: Magazines as a Source of Information

	Users (n=886) Percent	Non-Users (n=358) Percent
Unimportant	15.1	19.6
Neutral	35.4	40.5
Important	49.4	39.9
Total	100.0	100.0

(Chi-square = 9.7, d.f. = 2, p = 0.0.)

5.5.2 Interpersonal Sources

Respondents were asked if they thought friends, family members and relatives were important sources of information. A total of 68.3% users and 59.1% non-users considered friends as important. Meanwhile, 71.2% users and 64% non-users regarded family and relatives as crucial sources. Results are presented in Tables 5.5.6 to 5.5.7.

In both cases, there were more users than non-users who said that family and relatives, and friends were important sources. However, the reverse was true for adults. Overall, Secondary One students, regardless of users or non-users, seemed to regard family and relatives as important sources. By comparing Tables 5.5.1 to 5.5.5 and Tables 5.5.6 to 5.5.7, it becomes evident that interpersonal sources were important to and, in some cases, more important than certain mass media. Again, this may reflect some fundamental differences between students and adults in terms of their information needs and cognitive ability. In our subsequent surveys, it will be interesting to find out whether and what changes will take place as these youths mature and enter a different social environment.

Table 5.5.6: Friends as a Source of Information

	Users (n=883) Percent	Non-Users (n=357) Percent
Unimportant	6.3	10.6
Neutral	25.4	30.3
Important	68.3	59.1
Total	100.0	100.0

(Chi-square = 11.7, d.f. = 2, p = 0.0.)

Table 5.5.7: Family and Relatives as a Source of Information

	Users (n=884) Percent	Non-Users (n=356) Percent
Unimportant	7.6	11.2
Neutral	21.3	24.7
Important	71.2	64.0
Total	100.0	100.0

(Chi-square = 7.2, d.f. = 2, p = 0.0.)

5.6 Trust of Institutions

To further explore the relationship between Internet use and the media, respondents were asked if they trusted the news obtained from television, radio, newspapers and the Internet. Overall, the proportions of respondents who said they had trust in television, radio and newspaper news were high. About 78% of the users and 70% of the non-users said they trusted news from newspapers. This was followed by television news, which received support from 72.9% users and 69.9% non-users. Radio news came in third, with 63% users and 61% non-users saying they trusted it. Online news fared less well, yet it was still considered trustworthy by 58% users and 57% non-users. Breakdowns of the percentages are available in Tables 5.6.1 to 5.6.4.

This indicated that traditional news media were still credible in the eyes of many Secondary One students. The pattern resembled that of the adult respondents, who also trusted traditional news sources and were less ready to believe online news.

Table 5.6.1: Trust in Television News

	Users (n=884) Percent	Non-Users (n=359) Percent
Do not trust	5.8	8.4
Neutral	21.4	21.7
Trust	72.9	69.9
Total	100.0	100.0

(Chi-square test indicated no significant difference between users and non-users.)

Table 5.6.2: Trust in Radio News

	Users (n=884) Percent	Non-Users (n=359) Percent
Do not trust	8.1	12.0
Neutral	28.8	27.0
Trust	63.0	61.0
Total	100.0	100.0

(Chi-square test indicated no significant difference between users and non-users.)

Table 5.6.3: Trust in Newspapers' News

	Users (n=883) Percent	Non-Users (n=358) Percent
Do not trust	4.8	6.4
Neutral	17.7	23.7
Trust	77.6	69.8
Total	100.0	100.0

(Chi-square = 8.2, d.f. = 2, p = 0.0.)

Table 5.6.4: Trust in Online News

	Users (n=884) Percent	Non-Users (n=357) Percent
Do not trust	8.1	9.8
Neutral	33.7	33.3
Trust	58.1	56.9
Total	100.0	100.0

(Chi-square test indicated no significant difference between users and non-users.)

In addition to their view on the media, respondents were asked to indicate if they trusted the government, business corporations, and religious organisations. A total of 72.8% users and 70% non-users said they trusted the government, and only about 6% said they did not trust the government (see Table 5.6.5).

The business community and religious organisations, however, received far less trust from the students than the government did. (Refer to Tables 5.6.6 to 5.6.7.) A total of 32.8% users and 33.2% non-users said that businesses and corporations could be trusted. Similarly, 37.9% users and 39.2% non-users trusted religious organisations. However, it is important to note that 44.5% to 52% of the respondents expressed a neutral position towards the business community and religious organisations. Understandably, many of them felt they were not in a position to assess whether such institutions could be trusted.

Again, the findings were consistent with those from the 1999 adult survey, in which a larger proportion of respondents expressed trust in the government, compared with those in the business community and religious organisations.

Table 5.6.5: Trust in Government

	Users (n=883) Percent	Non-Users (n=357) Percent
Do not trust	6.1	5.6
Neutral	21.1	24.4
Trust	72.8	70.0
Total	100.0	100.0

(Chi-square test indicated no significant difference between users and non-users.)

Table 5.6.6: Trust in Business Companies/Corporations

	Users (n=884) Percent	Non-Users (n=358) Percent
Do not trust	15.0	14.5
Neutral	52.1	52.2
Trust	32.8	33.2
Total	100.0	100.0

(Chi-square test indicated no significant difference between users and non-users.)

Table 5.6.7: Trust in Religious Organisations

	Users (n=875) Percent	Non-Users (n=355) Percent
Do not trust	15.1	16.3
Neutral	47.0	44.5
Trust	37.9	39.2
Total	100.0	100.0

(Chi-square test indicated no significant difference between users and non-users.)

CHAPTER 6

Summary & Conclusion

Although there may be some immediate and short-term effects brought about by the introduction of a new medium, the understanding of social impact usually takes time. This monograph represents the first attempt of the Singapore Internet Project in examining the social impact of the Internet. From the massive amount of data collected, only the key findings from the first nation-wide survey are captured in this publication. They serve as benchmarks for tracking the influence of the Internet on Singaporeans over time.

6.1. Internet Access

The 1999 data shows that Singapore has a high level Internet penetration, with about 46% of the adults, age 18 and above, being Internet users. This is not far off from the Internet penetration of other countries such as Sweden and the United States, which stand at 56% and 59%, respectively (Nielson Net Ratings, 2001). The Internet penetration stands at 71% for the student sample, which is 25% points higher than that of the adult population. With the government advocating Internet use in schools and incorporation of information technology skills in the curriculum, Internet adoption among students is likely to increase over the years.

For the non-users, the top three reasons for not using or having stopped using are: did not know how, no time and no interest. These, of course, included those who have no chance to come in contact with the new medium, or it is not required for their work or study. For the latter category of non-users, the busy lifestyle of Singaporeans can be a major barrier to Internet adoption. Of equal concern are those who are lagging behind in acquiring IT skills. Others simply do not find the Internet to be useful or interesting to them.

In comparing the demographic profiles of Internet users and non-users, it can be seen that a digital divide exists and it is clearly linked to the social stratification in Singapore society. As compared to non-users, more Internet users are male, single and younger in age. Users tend to be of higher socio-economic status in terms of education, income and housing type. The majority of student and adult users know all the necessary skills in utilising the Internet, except that less than half of the users know how to set up a web page. This finding is not unique to Singapore. These characteristics are similar to the average Internet user in the United States as reported by the UCLA Internet Report (2000).

Among the adults, non-users were found to show a higher level of alienation than users. Users were less likely to feel helpless, think they could not rely on others and believe others did not care. The analysis is correlational in nature. Since users are generally of higher social-economic status, the negative relationship between alienation and Internet use could be spurious. No conclusion can be made at this point. The tracking of alienation scores over time will help us to see if alienation increases with Internet use. As for the students, no difference was found between users and non-users in their alienation scores — users did not experience a stronger sense of helplessness and isolation anymore than did the non-users.

As Singapore moves toward an IT- and knowledge-based society, the digital divide deserves serious attention. The problem of an ageing population, and the fact that older folks and people in the lower stratum are the laggards in the cyberspace race, all point towards the need for immediate remedial actions. Again, this problem is not unique to Singapore as several studies have found that the elderly are the ones that are least interested, have the least access and are, thus, the least likely to adopt this new technology (Lenhart, 2000; Harvard Kennedy School of Government, 2000).

Research has shown that accessibility to the use of a computer is an important factor in determining the level of web usage and, similarly, the utilisation of information technology (Novak and Hoffman, 1998; Goslee and Conte, 1998). Hence, one of the strategies by which the digital divide could be narrowed between the haves and the have-nots was to ensure adequate access to information technologies such as the personal computer and Internet facilities for those who cannot afford it (Novak & Hoffman, 1998). Such a strategy could be carried out via community-based projects, provided that funding and resources are made available (Goslee & Conte, 1998). More access points could be established in public institutions such as schools and libraries — places where those who lack personal computer access at home are likely to seek for connectivity (NTIA, 1999).

Unlike the adult population, there was no gender divide among the students. The data shows that males and females were equally likely to use the Internet. Again, the wide spread application of the Internet in many schools may account for its equal access among students of both genders. This, together with the high Internet penetration among students, is a result of the widespread introduction of computers to schools in Singapore. Although the digital divide in gender terms does not exist among the student population, the data shows that it exists along the lines of other socio-economic factors. Similar to the findings from the adult survey, the users in the student sample tend to come from families who were better off in terms of family income, housing type and parents' education, as compared with non-users.

However, if the penetration rate continues to grow among this group of young people, as we believe it will, the phenomenon of "digital divide" may be alleviated. The government is promoting IT and IT culture, beginning at the primary level in all schools in Singapore, with enormous investments in both human and financial resources. There is good reason to believe that educational institutions can serve an important "equalizing" function to reduce the "digital divide" and moderate the influence of family background in IT literacy and Internet use in school and in future career development. Obviously, more can still be done in Singapore at the school level.

6.2 Internet Activities

E-mail and information search are the two most popular Internet activities across the board, followed by entertainment and online discussion/chat, while e-commerce transactions are not common. On the average, users had about 2.7 years of Internet experience and spent about 11 hours a week on the Internet — of which about 3 hours were on e-mail, 2.6 hours on information search for work/school and 1.8 hours on information search for personal reasons.

On the average, students have been using the Internet for 1.6 years and they typically log on for about 5.4 hours per week. The Internet was also used for entertainment (1 hour a week), information search for schoolwork (1 hour a week) and e-mail (0.8 hour a week). Reading online news and conducting online transactions were not popular. Home use accounts for 61% (or 3.3 hours per week) of Internet time. Students also access the Internet from school (about an hour a week) and from other places, such as the library. The heavy usage at home highlights the importance of family and parents in monitoring young people's access to the Internet.

E-commerce on the other hand is not common. The main concerns about online transaction were security, privacy and the difficulty in assessing product quality. Although 45% of the Internet users had browsed for goods and services, less than 11% had actually made transactions. While 45.4% of users have taken advantage of the Internet to search for goods and services on online (i.e. browsers), only 10.6% of users have made purchases. Time should not be a decisive factor as browsers were willing to spend the time on information search, and the additional time to make the purchase should be quite insignificant. Skill was not the problem either because the data shows that purchasers and browsers have similar skill profiles.

The amount of online shopping and browsing for goods and services is likely to be higher in the future, given the overall level of Internet access in Singapore. E-commerce operators will find much potential in growth in catering to women shoppers and providing online educational materials and services.

Consumer concerns about security and privacy of information suggest that public perception of the security risks of e-commerce needs to be managed through education and promotion. Steps taken by the government and by businesses to address security risks need to be publicised. The traditional media, such as newspapers, and radio and TV broadcasts are potentially effective for this purpose. Our study shows that the level of trust in the traditional media is high in Singapore.

The other major concerns deal with assessing product quality and product delivery and returns. These reflect Singaporean consumers' lack of experience with catalogue shopping. Singapore businesses also need to learn to provide comprehensive product/service information and to provide strong support for product returns/exchanges and refunds.

Finally, consumers may also have fewer reasons to transact online given the small geographical size of Singapore and the efficient transportation infrastructure. Businesses are more likely to attract online purchasers when they provide comprehensive information, ease of product search and offer products/services not available elsewhere, and 24-hour availability.

6.3 Attitude and Perceptions

Both adult users and non-users have positive perceptions on the utility of the Internet, although Internet users show far more favourable attitudes toward the Internet than non-users. Both the adult users and non-users maintained positive perceptions of the Internet being "a good thing for society" and that "the Internet is useful to stay informed". Users differed from non-users in that they are more likely to see the Internet as "great for entertainment purposes", "makes business/commercial activities more convenient" and that it is "internet skills are important for getting a good job".

Student perceptions were similar to that of the adults with both users and non-users holding positive views toward the Internet, although the former tend to be even more positive in their perception. The vast majority of users thought that the Internet was important, useful, interesting, easy to use and convenient. Although the percentages of users who held favourable attitudes towards the Internet were higher than those of the non-users, this did not mean non-users were critical of the new medium. In fact, most of the non-users also shared the views of the users that the Internet was important, useful, interesting, easy and convenient to use. In addition, the non-users believed that the Internet was "a good thing for society", great for entertainment purposes" and that "Internet skills are important for getting a good job".

The results suggest that apprehension and perceived irrelevance can be psychological barriers for Internet adoption. While digesting these findings, one should bear in mind that adult non-users are older and lower in socio-economic standing. Removing the psychological barriers would be a good starting point if the government wants to help the non-users enter the Internet era. The positive perception may be partly attributed to the government's persuasive messages on the importance of IT and skilled workers for the development of society, as well as frequent media coverage of the Internet. However, it should be noted that only 62% of users and 39% of non-users thought that Internet skills are important for getting a good job (the lowest percentage agreement among the 5 perception items for both users and non-users). This is an area the government may want to look into in its push for an intelligent island.

6.4 Lifestyle and Well-being

There were significant differences between the adult users and non-users in terms of the time that they spent on media and leisure activities, with the exception of the time spent reading newspapers and listening to the radio. Generally, users spent more time reading books, playing video/PC games, listening to recorded music, reading magazines, talking on the phone, going to the cinema, participating in sports and interacting with friends. Non-users spent more time watching television and interacting with family members than users. This might be due to the fact that, compared with users, non-users were generally less educated, older and married, which might explain why they spent more time watching television and interacting at home with the family.

Statistical tests showed no significant differences between student users and non-users in the time they spent on these activities. Perhaps this is because the daily activities of the 12 to 13-year-olds are often scheduled by the adults around them, either at home or in school. As such they are not free to follow their own desires and interests just yet, and, therefore, there is no significant difference between users and non-users on the media and leisure activities.

Taken as a whole, the Internet is considered beneficial to many people and it seems to have positive influences on cultivating friendships. The data collected from this study does not lend support to the fear that the Internet will erode existing friendship and familial ties. Almost half of the adult users said they have made more friends with the use of the Internet, while only 13% reported having fewer friends. As for the student users, about 55% reported having more friends and 36% said they spent more time with close friends after they started using the Internet. Users also reported having more contact with people who shared their hobbies (41%) and professions (45%), far exceeding those who reported having less contact (17% and 8%, respectively). The proportions of students who indicated no change in making friends and time spent with friends were also considerably large, at 37.5% and 47.7%, respectively.

For many people, cyberspace is just another meeting place and it has increasingly been seen as a medium for social interaction. The findings from this study give credence to the suggestion put forth by Katz and Aspden (1997) that the Internet is indeed a place where friendships can develop. This point is especially salient when applied to the student population. These students are still fairly young and hence for most of them, besides attending lessons in school and participating in school co-curricular activities, they would be expected to return home immediately after school. The use of chatrooms and emails would offer an alternative for them to socialise with their friends in the comfort of their own homes. Using the Internet was a way that they could maintain contact with their friends, especially via email and Internet Relay Chat (IRC). This might account for the fact that about one-third of them said that they had spent more time with their close friends. Other scholars have found that the use of the Internet for communication has the potential to help cement pre-existing offline bonds rather than harming users' offline social network ties (Hamman, 1988). In addition, relationships that begin online rarely stay online and often result in face-to-face meetings (Parks & Floyd, 1996; Wellman & Hampton, 1999).

There was no indication that Internet use had led to noticeable negative effect on family life. A comparison between users and non-users showed little difference in terms of family communication pattern. Half or more of the adults from both groups reported having relatively positive experiences with members of their family. They talked to each other, listened to each other, shared ideas and opinions, and were patient with each other. Overall, there was no difference between the two groups in terms of the amount of time spent as well as the quality of family communication. Most Internet users (about 60%) report no change in time spent with family and close friends. The difference between those who reported spending less time (24.2%) and those spending more time (17.1%) is not significant. This result was almost identical for the student users.

Internet use has not led to a widespread increase in the practice of bringing work home and working from home. Only about one in ten of the adult users who were working said that with the use of the Internet, they brought work home to do at night and they did this approximately three days a week. Furthermore, the same percentage of users said that Internet use had allowed them to work from home. The figures are similar for those who work from home during office hours. The monitoring of these figures over the next few years will enable us to determine if the Internet is allowing work to eat into family time or if it actually enhances flexibility in telecommuting. This has important implications on work, family life and parenting. In fact, Internet use is equally common at home and at work, and is reported to increase productivity.

6.5 Media use

There seems to be no immediate concern that the Internet is replacing traditional mass media. Some studies suggest that Internet users tend to be heavy media users in general, and that the Internet has not replaced the traditional sources (Bromley & Bowles, 1995; Kaye, 1998). The data from this study reveals that Internet users are exposed to all kinds of media, more so than non-users. A great majority of the users continue to view newspapers, television and radio as important sources of information. Notably, seven in ten of the users also think the Internet is an important source of information. The pattern was similar for non-users, except that in their case, the Internet came in last. This is similar to the finding by the Pew Research Centre (1998), which revealed that going online for news did not affect news consumption patterns of traditional news sources. The Internet appears to act as a complement to the other media rather than act as a form of competition or a substitute. According to Stemp III, Hargrove and Bernt (2000), the new medium did not cause the decline in use of other media, and neither was it a strong competitor to the other media. In fact, Internet users are more likely to consume other media concurrently as compared to non-Internet users.

A higher percentage of users than non-users identify the mass media as important sources of information. However, the reverse seems to be true when it comes to interpersonal sources. For the non-users, 72% consider friends as an important source and 80% see family members and relatives as important sources. The figures for users are 65% and 66%, respectively.

In this study, it was found that the trust levels of news from traditional mass media remain high. A large proportion (about three-quarters) of both adult users and non-users said that they trust the news from television, radio and newspapers. When it comes to online news, only 47% of the non-users find it trustworthy, as compared to 60% of the users. A significant percentage of non-users remain "neutral" as they may not be in a position to comment on trustworthiness of online news. Research on traditional media suggests that credibility is strongly related to the frequency with which individuals use that particular medium (Wanta & Hu, 1994; Westley & Severin, 1964; Carter & Greenberg, 1965). This finding might not be applicable to the Internet because the Web is an emerging medium, so levels of experience may not yet influence judgments of credibility.

The Internet is still a novice to many people despite the high penetration rate. Hence, it is not surprising for online news to lag behind the traditional media in terms of credibility. Past research also suggests that reliance is more strongly associated with media credibility than with general use measures (Wanta & Hu, 1994; Rimmer & Weaver, 198). Thus, in order for online news sources to increase their readership, they would need to increase their credibility so as to gain the trust of the people.

Results from the student survey are similar with about 78% of users considering the Internet as an important source of information. Meanwhile, newspapers are a vital source to 70% users and 62% non-users. Television is also considered important to 66% users and 54% non-users. Has the Internet established itself as the most important source and threatened the position of the traditional media among young Singaporeans? The answer is not conclusive, as the sample of the present survey consisted only of the 12 to 13-year-olds. For this group, it is not that traditional media have been replaced — it is likely that they have never been used extensively to begin with. After all, the respondents are young — they may not have the need or the cognitive ability to use these traditional media. In fact, the study shows that among this group of youths, both users and non-users rely on interpersonal sources, such as family members, relatives and friends, for information. While they are different from the adults in many ways, we do not know to what extent this new "Internet" generation will continue to be different from the "pre-Internet" generation when they grow up. Subsequent surveys of this same group of respondents will provide relevant data to shed some light on the changing patterns of Internet usage and its impact on the individual and society.

The students' perception of trust toward the various media is similar to that of the adult population. Thus, another possible explanation is that young people are acquiring these attitudes from older people around them, such as parents, teachers and older siblings. One possible explanation is that traditional media have been around longer and have established their credibility. They are properly licensed and the information they provide is collected and disseminated by professionals. This is not necessarily the case with the news coming from the Internet, where anyone with technical skills can put up any information (or misinformation) that resembles news. In addition, the assumed strength of the Internet, i.e. it is a freewheeling, unregulated outpost for anyone to express his or her opinions, might also weaken the credibility value. There are no professional and societal pressures to provide accurate and unbiased information, unlike the traditional news sources. There is also the presence of parody sites that might mislead users into mistaking them for official sites.

6.6 Internet Regulation

The easy availability of online pornographic material is a concern to the users (64%), but just over half of the non-users said that they were concerned about online pornographic materials. Similarly, more users than non-users expressed concerns over undesirable racial, political, and religious content.

Internet regulation, a subject of much interest to the public, the government, the content providers and the service providers alike, will continue to be a challenge. The benefits of the Internet are clear and compelling enough for wide spread adoption, but at the same time there is a concern that people may be affected by content that is deemed to be undesirable. More so than any other medium, with the Internet, the fundamental dilemma is between excessive control to the detriment of the users and excessive freedom to the detriment of society. However, the ever-changing technology behind Internet content and delivery systems may oblige ever-increasing self-regulation by users.

The findings of this survey suggest that the current "light touch" regulatory framework adopted by the government is consistent with public desire. The Singapore government has already begun to implement what is called a "balanced and light-touch" regulatory framework. The government has stated that it was designed to take advantage of the full potential of the Internet while, at the same time, maintain social values and racial and religious harmony in Singapore. The government has also openly declared that it aims for minimal legislation and wishes to encourage more industry self-regulation and public education so that users are empowered to use the Internet for its benefits.

For the public too, there are some serious implications. It is not at all clear from this survey (nor was it sought) how much the respondents really understood the technical limitations of Internet regulation. There was a popular belief that the Internet could be effectively controlled by the government and/or the ISPs. There is perhaps a residual effect from the previous regulatory regime in which not only did the government wish to but was also able to exercise unilateral control with great success. Given the reality that forced control in Singapore's telecommunication environment is becoming ineffective, it would be interesting to see how the public might respond to the call — by the government as well as by academic commentators — to build a social immune system against undesirable content. The extraordinarily high percentage (92%) of respondents who mentioned that individuals should be responsible, and about a two-third majority of respondents who indicated that the ISPs, government, central body as well as family should be responsible for reducing or avoiding undesirable Internet content reflect the public desire for concerted efforts from all parties involved, from the individual to the government. Given that the higher educated segment is more receptive to such facilities, it is important to observe whether this implies that a vocal elite may emerge to apply increasing pressure for such provisions.

6.7 Political Empowerment

The Internet has not led to a feeling of a stronger sense of political empowerment. Users of the Internet do not feel that the Internet can enable people to have more political power, more say about what the government does, better understand politics and make public officials care more about what people think. Neither has it brought significant change in terms of increasing contact with people who share their political interests and religion. The percentages of users who think that the Internet can enable people to have more political power and to have more say about what the government does are low (less than 20%). A large portion of users, however, believe that the Internet allows them to understand politics better and that public officials will care more about what people think (40% and 31%, respectively).

6.8 Trust of Institutions

Trust in the government was high for both adults and students, but business and religious institutions did not fare that well. Both adult users and non-users hold a great deal of trust in the government (about 70%), but only 39% to 45% trusted business corporations. Religious organisations managed to win the trust of only 42% of the both users and non-users. However, not many people distrusted these institutions. Instead, a large percentage remained neutral (40% to 43%) with regard to both institutions.

Most students, like adults, are ready to trust the government, but not many indicated such trust in the business community and religious institutions. About 70% said they trusted the government while only about 33% to 39% had trust in business and religious institutions. There were no significant differences between users and non-users. We observe smaller percentages of students who express trust in them and a higher proportion of neutrals. One explanation is that students have limited contact with business and religious institutions, which are of little relevance to them. The expressed trust in the government, on the other hand, reflects the prevailing perception in Singapore that students learn from adults. It is obvious that students in Singapore, like their elders, are mostly pro-establishment. We are yet to find out whether continuous exposure to the Internet from a young age may influence this pro-establishment attitude in the long run.

6.9 Looking Ahead

Overall, this survey finds that the Internet penetration rate among both adults and students is relatively high at 46% and 71%, respectively. With the Ministry of Education's active promotion of Information Technology and their drive to increasingly incorporate it into the curriculum, we expect to see more non-users among the student population picking up Internet skills over the next few years. Differences in socio-economic background will play a less significant role in determining whether or not students become Internet users in the future. This will result in greater Internet access parity as the future generation embraces this new technology.

Both the adult and the student populations, and both users and non-users, generally share positive perception of the Internet. This provides a strong foundation for further utilisation of the Internet by existing users and would possibly push many non-users to decide to adopt this new technology. By comparing users with non-users, and based on the respondents' self-reports, we have not discovered any indication that Internet use is leading to undesirable consequences. In fact, many express a positive view that the use of the Internet will bring benefits to its users and society in general. It should be noted that these findings are based on self-reports, which may or may not reflect reality. As mentioned repeatedly in this monograph, many of the indicators are meant to be baseline measurements. We will have to continue to monitor the situation in subsequent surveys to establish a better understanding of the impact of the Internet among the Singapore population.

We can expect to have a highly computer literate younger generation of Singaporeans in the near future. As for the older Singaporean, the government has started several programmes to encourage and educate the population to adopt the use of computers and Internet into their everyday lives. Public access to computers and the Internet is also readily available in the public libraries as well as in numerous cybercafes. All these measures will help to narrow the digital divide. However, as the problem of adoption and access to this new technology is rapidly being eradicated, a new one is in sight. Much of the content on the Internet appears to be relatively difficult to understand with a readability level that is beyond that of the less educated and aged populace. Thus, the digital divide along different lines might persist. This time it will be along the lines of "content" rather than "access", with the purpose of going online being the divisive factor. The knowledge gap within society might be increased as individuals go online with different intentions (e.g. searching for information vs. going online to play games). However, research in this area is still lacking at this point in time and much still needs to be done in this area.

References

Asociacion para la Investigacion de los Medios de Comunicacion (July 13, 2001). *Comprehensive study of Spanish Net use.* Retrieved Aug 2, 2001 from the World Wide Web: www.nua.ie/surveys/index.cgi?f=VS&art_id=905356974&rel=true

Athaus, S. L., & Tewksbury, D. (2000). Patterns of Internet and traditional news media use in a networked community. *Political Communication,* 17, 21-45.

Australian Bureau of Statistics (May 11, 2001). *A third of Australian homes now online.* Retrieved August 02, 2001 from the World Wide Web: www.nua.ie/surveys/index.cgi?f=VS&art_id=905356752&rel=true

Beninger, J. (1987) Personalization of mass media and the growth of pseudo-community. *Communication Research,* 14, 352 – 371.

Berteismann Foundation Germany & Australian Broadcasting Authority (1999). Risk assessment and opinions concerning the control of misuse on the Internet. In J. Waltermann & M. Machill (Eds.), *Protecting our children on the Internet. Towards a new culture of responsibility.* Berlesmann Foundation Publishers.

Canadian News Wire (June 2, 2000). *Internet Faces Credibility Gap: Canadian Survey.* Retrieved Sept 1, 2001 from the World Wide Web: http://www.newswire.ca/releases/June2000/02/c0971.html

Calhoun, C. (1991) 'Indirect Relationships and Imagined Communities: Large Scale Social Integration and the Transformation of Everyday Life', in J. Coleman and P. Bourdieu (eds) Social Theory for a Changing Society. Boulder, CO and San Francisco: Westview Press.

Carter, R., & Greenberg, B. (1965) Newspaper or television: Which do you believe? *Journalism Quarterly, 42(4),* 29 – 34.

Chesebro, J., & Bonsall, D. (1989). *Computer-mediated Communication : Human relationships on a computerized world*. Tuscaloosa: University of Alabama Press.

Chueng, W., Chang, M. K., & Lai, V. S. (2000). Prediction of Internet and World Wide Web usage at work: A test of an extended Triandis model. *Decision Support Systems, 30*, 83-100.

Hindman, D. C. (2000). The rural-urban digital divide. *Journalism & Mass Communication Quarterly, 77*(3), 549-560.

Garton, L., & Wellman, B. (1995). Social impacts of electronic mail in organizations: A review of the research literature. *Communication Yearbook, 18*, 434-453.

Goslee, S., & Conte, C. (1998). *Losing ground bit by bit: Low income communities in the information age*. Benton Foundation's Communications Policy and Practice Program. Retrieved Jan 12, 2001 from World Wide Web: http://www.benton.org/Library/Low-Income/

Hamman, R., (1998). *The online/offline dichotomy: Debunking some myths about AOL users and the effects of their being online upon offline friendships and offline Community*. MPhil Thesis, University of Liverpool, Dept of Communication Studies. [Online] Available: http://www.socio.demon.co.uk/mphil/index.html (Sept 1, 2001)

Hampton, K.N., & Wellman, B. (1999). Netville online and offline – Observing and surveying a wired suburb. *American Behavioral Scientist, 43*(3), 475-492.

Hampton, K.N., & Wellman, B. (2000). Examining community in the digital neighbourhood: Early results from Canada's wired suburb. In T. Ishida & K. Isbister (Eds.), *Digital cities: Technologies, experiences, and future perspectives* (pp. 194-208). Lecture Notes in Computer Science 1765. Heidelberg, Germany: Springer-Verlag.

Harvard Kennedy School of Government (2000). *Technology survey for adults and kids*. Retrieved Jan 1, 2001 from the World Wide Web: www.npr.org.programs/specials/poll/technology/technology.kids.html.

Haythomthwaite, C., & Wellman, B. (1998). Work, friendship, and media user for information exchange in a networked organisation. *Journal of the American Society for Information Science, 49*(12), 1101-1114.

Hoffman, D. L., & Novak, T.P. (1999). *The evolution of the digital divide: Examining the relationship of race to Internet access and usage over time*. Retrieved Jan 12, 2001 from the World Wide Web: www2000.ogsm.vauderbilt.edu.

IMRBINT (2001). *How Many Online?* Retrieved Aug 2, 2001 from the World Wide Web: www.nua.ie/surveys/how_many_online/asia.html

ITU (2001). *How Many Online?* Retrieved Aug 2, 2001 from the World Wide Webs: www.nua.ie/surveys/how_many_online/asia.html

Iivari, J., & Igbaria, M. (1997). Determinants of user participation: A Finnish survey. *Behaviour and Information Technology, 16(2)*, 111-121.

James, M. L., Wotring, E., & Forrest, E. J. (1995). An exploratory study of the perceived benefits of electronic bulletin board use and their impact on other communication activities. *Journal of Broadcasting and Electronic Media, 39,* 39-50.

Jupiter Media Metrix (Apr 18, 2001). Internet users in US spending longer online. Retrieved Aug 2, 2001 from the World Wide Web: www.nua.ie/surveys/index.cgi?fVS&art_id=905356672&rel=true

Katz, J.E., & Aspden, P. (1996). *Friendship formation in cyberspace : Analysis of a national survey of users.* [Online] Available : http://www.nicoladoering.net/Hogrefe/katz.htm

Kiesler, S., Siegel, J., & McGuire, T.W. (1984) Social Psychological Aspects of Computer-Mediated Communication. *American Psychologist, 39(10),* 1123-1134.

Koku, E., Nazer, N., & Wellman, B. (2000). *Netting the scholars: Online and offline.* Retrieved Aug 2, 2001 from World Wide Web: http://www.chass.utoronto.ca/~wellman/publications/index.html.

Korean Ministry of Information and Communication (Apr 20, 2001). *Half of Koreans now regular Net users.* Retrieved Aug 2, 2001 from the World Wide Web: www.nua.ie/surveys/index.cgi?f=VS&art_id=905356679&rel=true

Kraut, R., & Lundmark, V. (1998). Internet paradox: A social technology that reduces social involvement and psychological well-being. *American Psychology, 53(9),* 1017-1031.

Kraut, R., Mukhopadhyaya, T., Szczppula, J., Kiesler, S., & Scherlis, B. (1999). Information and communication: Alternative uses of the Internet in households. *Information Systems Research, 10(4),* 287-303.

Kraut, R., Mukhopadhyaya, T., Szczppula, J., Kiesler, S., & Scherlis, B. (1996). *Communication and Information: Alternative Users of the Internet in Household.* Retrieved Jan 1, 2001 from the World Wide Web: www.acm.org/owsbin/dl/ows/dl.rearch

Larsen, E. (2000). *Wired churches, wired temples; taking congregations and missions into cyberspace.* The Pew Internet & American Life Project.

Lee, T., & Birch, D. (2000). Internet regulation in Singapore: A policying discourse. *Media International Australia Culture and Policy: International Issues in Media Regulation, 95,* 147-169.

Lenhart, A. (2000). Who's not online: 57% of those without Internet access say they do not plan to log on. *Pew Internet & American Life Project.* Retrieved Jan 1, 2001 from the World Wide Web: www.pewinternet.org

Maignan, I., & Lukas, B. A. (1997). The nature and social uses of the Internet: A qualitative investigation. *The Journal of Consumer Affairs, 31(2),* 346-371.

National Telecommunications and Information Administration: US Department of Commerce (1999). *Falling Through the Net: Defining the Digital Divide.* Retrieved Jan 1, 2001 from the World Wide Web: www.ntia.doc.gov

Newsbyte Asia (2001). *How Many Online?* Retrieved Aug 2, 2001 from the World Wide Webs: www.nua.ie/surveys/how_many_online/asia.html

Nielsen NetRatings (2001). *How Many Online?* Retrieved Aug 2, 2001 from the World Wide Webs: www.nua.ie/surveys/how_many_online/asia.html

Novak, T.P., & Hoffman, D. L. (1998). *Bridging the digital divide: The impact of race on computer access and Internet use.* Owen Graduate School of Management, Vanderbilt University, Nashville, TN 370230. Retrieved http://www2000.ogsm.vanderbilt.edu/

Parks, M., & Floyd, K (1996) Making Friends in Cyberspace. *Journal of Communication. (46)1,* 80 – 84

Pew Research Centre (1998), *Study on Media Consumption.* Retrieved Sept 1, 2001 from World Wide Web: http://www.people-press.org/med98rpt.htm

Rainie, L., & Packel, D. (2001). *The Pew Internet & American Life Project.* Retrieved Jan 1, 2001 from the World Wide Web: www.pewinternet.org

Rimmer, T., & Weaver, D. (1987). Different questions, different answers? *Journalism Quarterly, 64,* 28-36,44.

Romm, C. T., & Pliskin, N. (1999). The office tyrant – social control through e-mail. *Information Technology and People, 12(1),* 27-43.

Spink, A., Bateman, J., & Jansen, B. J. (1999). Searching the Web: a survey of EXCITE users. *Internet Research: Electronic Networking Applications and Policy, 9(2),* 117-128.

Stempel III, G. H., Hargrove, T., & Bernt, J.P. (2000). Relation of growth of use of the Internet to changes in media use from 1995 to 1999. *Journalism & Mass Communication Quarterly, 77(1)*, 71-79.

UCLA Centre for Communication Policy. *The UCLA Internet Report: Surveying the Digital Future*. Retrieved Oct 30, 2000 from World Wide Web: www.ccp.ecla.edu.

Venkatesh, A. (1996). Computers and other interactive technologies for the home. *Communications of the ACM, 39(12)*, 47-54.

Wellman, B. (1996). An electronic group is virtually a social network. In S. Kiesler (Ed.), *Cultures of the Internet* (pp. 179-205). Hillsdale, NJ: Lawrence Erlbaum.

Wellman, B. (2000a). Changing connectivity: A future history of Y2.03K. *Sociological Research Online, 4(4)*. Retrieved Jan 1, 2001 from the World Wide Web: www.socresonline.org.uk/4/4/wellman.html

Wellman, B. (2000b). *Physical place and cyberplace: The rise of networked individualism*. Retrieved Jan 1, 2001 from the World Wide Web: www.chass.utoronto.ca/~wellman/publications/individualism/article.html

Wellman, B., & Hampton, K. (1999). Living networked in a wired world. *Contemporary Sociology, 28(6)*. [Online] Available: http://www.chass.utoronto.ca/~wellman/publications/index.html (2001, July 31).

Wellman, B., Salaff, J., Dimnitrova, D., Garton, L., Haythornthwaite, C., & Gulia, M. (1996). Computer networks as social networks: Collaborative work, telework & virtual community. *Annual Review of Sociology, 22*, 213-238.

Westley, B.H., & Severin, W.J. (1964) A Profile of the Daily Newspaper Non-reader. *Journalism Quarterly, 41(4)*, 45 – 50.

Wyden, R. (2000). *Oregon seniors and the digital divide – A survey of senior centers' Internet access in the new millennium*. Retrieved Jan 12, 2001 from World Wide Web:

http://www.benton.org/Library/Low-Income/

Zhang, K., & Hao, X. (1999). The Internet and the ethnic press: A study of electronic Chinese publication. *The Information Society, 15*, 21-30.

Acknowledgement

We are grateful for the research grant provided by the Infocomm Development Authority of Singapore (IDA) and the Singapore Broadcasting Authority (SBA). We are deeply touched by their trust of the research team and unfailing support for the project.

For the respondents who took part in our adult survey, we are grateful for their cooperation. Their voluntary participation provided crucial information on the usage and impact of the Internet in Singapore.

The student survey would not have been possible without the support and assistance from the Ministry of Education and Principals, teachers and students of the following randomly selected schools:

- Bartley Secondary School
- Bendemeer Secondary School
- Catholic High School
- CHIJ Secondary (Toa Payoh)
- Chong Boon Secondary School
- Crescent Girls' School
- Hong Kah Secondary School
- Junyuan Secondary School
- Jurongville Secondary School
- Methodist Girls' School
- Shuqun Secondary School
- St. Hilda's Secondary School
- St. Margaret's Secondary School
- Tanglin Secondary School

- Tanjong Katong Secondary School
- Thomson Secondary School
- Woodsville Secondary School
- Yio Chu Kang Secondary School
- Yishun Secondary School

We would also like to express our appreciation to the UCLA Center for Communication Policy for jointly embarking on this ambitious World Internet Project (WIP) with us. The support and participation from collaborating research teams all over the world has been most encouraging.

Last but not least, we would like to thank Miss Jolene Tan, Miss Loh Soo Fun and Miss Tracy Loh for their efficient research assistance in the project.

Index

access to the Internet, 7, 100-2, 113
activities, Internet, 29-31, 37-8, 78-9, 80, 103-4
adult survey, 22-69
advantages, Internet, 2, 13
age, 7, 8, 15, 16-17, 26-7, 33, 58-9, 77, 101
    sample vs. population, 22-3, 71
alienation, 2, 12, 43-6, 69, 84-7, 101
Asia, 7
Australia, 7, 8
business, trust in, 68, 98, 99, 111-12
censorship, 55-6, 58-9
chat rooms, 17, 30-1, 34, 77-8, 79, 103, 106
China, 5
Chinese students, 73-4
communities, Internet, 15
computer users, 25, 72
computer-mediated communication (CMC), 11-15, 16
content, 113
    see also regulation; undesirable content
demographic profiles
    adult survey, 22-4
    e-commerce, 33-4
    non-users, 25-9
    student survey, 70-7
    users, 7, 8, 25-9
digital divide, 15-17, 73, 77, 101, 113
disadvantages, Internet, 2-3, 12
e-commerce, 2, 17, 21, 31-6, 50, 91, 103
education level, 7, 15, 16, 22, 29, 33-4, 58-9, 65, 76-7, 101, 111, 113
    sample vs. population, 22, 24
education, IT integration, 1
elderly people, 17, 58, 59, 101, 113
emailing, 9, 12, 13, 14-15, 29, 30-1, 34, 77-8, 103, 106
employment prospects, 49, 51, 89, 91, 104, 105
entertainment, 9, 17, 50, 79, 80, 91, 103, 104, 105
Europe, 4, 32
family relationships, 12, 13, 38-42, 81-3, 95-6, 105-7, 108, 109

filter options, 58-9
Finland, 8
France, 5
future, 112-13
gender, 7, 8, 15, 17, 22, 23, 25-6, 33, 72-3, 102
    sample vs. population, 22-3, 71
government
    regulation, 18, 110
    trust in, 68, 98, 99, 111-12
home usage, 30-1, 43, 73-4, 107
Hong Kong, 5, 8
housing, 7, 28-9, 65, 75-6, 101
    sample vs. population, 22, 24
Hungary, 5
impact, 2-3
    see also alienation; family relationships; social relationships
implications of study
    government, ISPs, content providers and the public, 60
    institutions, 69-8
    IT promoters and policy makers, 53-4
    mass media and interpersonal communication, 65-6
income level, users, 7, 8, 15, 16, 27-8, 33-4, 65, 75
India, 5, 7
Indian students, 73-4
Indonesia, 7
inequalities
    see digital divide
information sources, 8-9, 21, 49, 60-6, 92-6, 107-8, 109
International Association of Communication (ICA), Conference, 5
International Association of Media and Communication Research (IAMCR) General Assembly and Scientific Conference, 5
Internet penetration rate, 24, 100, 112
Internet Service Providers (ISPs), 56-9, 110-11
IT 2000 Report, 1

Italy, 5
Japan, 5
jobs
  see employment prospects; work use
Korea, 5, 8
lifestyle, 20-1, 36-46, 80-7
location, Internet logon, 30-1, 78-9, 103
magazines, 61, 63, 92, 95
Malay students, 73-4
marital status, 27, 33, 65, 101
mass media, 9-11, 60-1, 92-5, 104, 107-9
media use, traditional, 9-10
  see also information sources; magazines; newspapers; radio; television
Ministry of Education, 1-2, 112
motivational factors, 8, 9
National Telecommunications and Information Administration, 15
New Zealand, 8
news
  credibility of media sources, 10-11, 66-7, 69, 97-8, 104, 108, 109
newspapers, 61, 62, 66, 67, 92, 94, 96, 109
non-use, reasons, 100-1, 105
North America, 4
Parents Advisory Group for the Internet (PAGi), 3
perception, Internet usefulness, 21, 46-54, 87-91, 104-5, 106, 109, 112
Pew Internet and American Life Project, 7
politics, 51-3, 54, 55, 69, 111
pornography, 2, 17, 55, 110
privacy, 17, 36, 103-4
productivity, 2, 42-3, 84
purchasing online, 31-3, 35-6
  see also e-commerce
purposes, 8-9
race, 7, 15, 16, 17, 18, 56, 73-4
  sample vs. population, 23
radio, 61, 62, 66, 67, 92, 94, 96
rating system, 57-8
recommendations, 102, 105
regulation, 3, 17-18, 21, 54-60, 110-11
religion/religious organisations, 38-9, 56, 68, 69, 98, 99, 111-12
rural areas, 7-8
Russia, 7
school type, student survey, 71
school use, 42-3, 74, 78, 84, 100, 102
security, 17-18, 36, 103

self-regulation, 18, 110
shopping online, 31-3, 35-6
  see also e-commerce
skills, Internet, 29-30, 34, 77-8, 101
social disadvantage
  see digital divide
social relationships, 11-15, 38-42, 81-2, 95-6, 105-6, 108, 109
  negative impacts, 12
  positive impact, 13
  sources of information, 63-5
socio-economic disparities
  see digital divide
student survey, 70-99
surfing, 17
surveys
  adult, 19, 20, 22-69
  establishment, 3-4
  focus, 4
  future, 113
  methodology, 19-21
  sampling, 19, 22-4
  student, 19, 20, 70-99
Sweden, 5, 7, 100
Taiwan, 5, 8
television, 9-10, 61, 62, 66, 92, 94, 96, 109
Thailand, 7
time spent online, 7, 11, 31, 37-8, 78, 79, 80
trust
  institutions, 66-9, 96-9, 111-12
  media, 10-11, 66-7, 69, 97-8, 104, 108
undesirable content, 2, 17, 21, 54-60, 110-11
  responsibility, 56-7, 110-11
  see also regulation
United States, 3, 7, 8, 10, 16, 17, 32, 34, 100, 101
usage patterns, 6-11, 20, 24-31, 71-9
users, number worldwide, 7
web site, setting up, 29, 34, 77-8, 101
well-being, 43-6, 80-7, 105-7
  see also activities; alienation
women, 8, 17, 103
  see also gender
work use, 42-3, 74, 78, 107
World Internet Project (WIP), 4-5
World Internet Project Conference, 5
young people, 17, 26, 58, 65, 70-99